T0301700

DIGITAL FILTERS

IEEE PRESS SERIES ON DIGITAL AND MOBILE COMMUNICATION

John B. Anderson, *Series Editor*
University of Lund

DIGITAL FILTERS
Principles and Applications with MATLAB

FRED J. TAYLOR

 John B. Anderson, *Series Editor*

IEEE PRESS

A JOHN WILEY & SONS, INC., PUBLICATION

Library of Congress Cataloging-in-Publication Data:
Taylor, Fred J., 1940-
  Digital filters : principles and applications with MATLAB / Fred J. Taylor.—1st ed.
     p. cm.—(IEEE series on digital & mobile communication ; 30)
  Includes index.
     ISBN 978-0-470-77039-9 (hardback)
  1. Electric filters, Digital.   2. Signal processing–Digital techniques–Mathematics.   3. MATLAB.
I. Title.
  TK7872.F5T39 2012
  621.3815'324–dc23
                                                                                2011021434

Printed in the United States of America
oBook ISBN: 9781118141151
ePDF ISBN: 9781118141120
ePub ISBN: 9781118141144
eMobi ISBN: 9781118141137

10 9 8 7 6 5 4 3 2 1

To my angel, Lori

CONTENTS

PREFACE

The history of digital filters essentially began in the mid-1970s, concurrent with the advent of the field of study called digital signal processing (DSP). Over the ensuing 30 something years, digital filters have become both a facilitating and enabling technology. They serve as analog replacements as well as serving in unique DSP roles in a host of application domains including communications, control, defense, audio, biomedicine, geophysics, radar, entertainment, and others. I have been blessed to be able to witness and participate in all of these phases of digital filter evolution.

A digital filter is a device that can modify the attributes of a signal using digital means. Required filter attributes can be assumed or defined in terms of published standards that specify amplitude and phase behavior as a function of frequency. Besides altering a signal's attributes, digital filters must often meet a host of other constraints such as speed, complexity, power consumption, cost, and other factors. In the pantheon of digital filters, the majority are identified as being finite impulse response (FIR), infinite impulse response (IIR), or multirate systems. The book's primary goal is to provide the needed understanding of both design and analysis strategies as they apply to mainstream digital filters.

In the normal course of an engineer's career, regardless of their disciplinary training, they will be called upon to design or analyze a mainstream filter. Unfortunately, many engineers and technologists have little to no formal digital filter experience. Fortunately, today's workplace is abundant with filter design software packages with various levels of sophistication. One of the leaders in this field is Mathwork's MATLAB™. Today, both practicing engineers and students of engineering exhibit a growing reliance on these tools with MATLAB being a de facto standard. However, after observing how these tools are being used in the workplace and classroom, concerns arise in that users are often overwhelmed with a plethora of filter design options, often developing a filter solution that may not be best for the target application. In addition, users often have insufficient experience or understanding of filter theory to be able to make even minor enhancements to a MATLAB-produced filter outcome. This too is a motivation for developing this book, which elevates the reader's understanding of how to characterize a digital filter, to make proper design choices, and to enhance a computer-generated design into a well-crafted outcome.

In reality, using tools such as MATLAB to design a mainstream digital filter is the easiest step in a solution process that ends with a successfully implemented digital filter. Implementation, whether in software or hardware, is generally the more challenging problem. Tools, such as MATLAB, provide the user with some basic implementation support. Unfortunately, most engineers have no, or only a rudimentary, understanding of the implementation choices offered by MATLAB. This provides additional motivation to develop filter implementation awareness skills, providing content that is generally missing in the current collection of digital filter books and monographs.

The book has been organized to support the stated objectives. The presentation begins with the fundamentals, including sampling, data acquisition, data conversion and quantization, and transforms. Next, the design, implementation, and analysis of an FIR filter are presented. Topics include FIR attributes, types, special cases, and implementation. Following FIRs, the design, implementation, and analysis of an IIR filter are presented. Like FIRs, topics include IIR attributes, types, special cases, and implementation. Additional attention is given to understanding state variables as an IIR architectural description language. Finally, multirate systems are explored, ranging from a discussion of their properties to case studies. In most cases, each topic is supported with MATLAB examples and exhibits

The study of filters is supported with a number of examples, many involving the use of MATLAB. In an attempt to actively engage the reader, the MATLAB script used to generate the MATLAB examples and graphics are available from John Wiley & Sons Supplemental Book Material site at http://booksupport.wiley.com. The MATLAB scripts can be easily copied into MATLAB's Command Window and reparameterized to reflect the reader's filter applications and needs. Many of the scripts were polished by Mr. Rajneesh Bansal, to whom I owe a great debt.

FRED J. TAYLOR
IEEE Fellow
Professor Emeritus, University of Florida
Board Chairman and Senior Scientist, The Athena Group Inc.

CHAPTER 1

INTRODUCTION TO DIGITAL SIGNAL PROCESSING

INTRODUCTION

Signal processing refers to the art and science of creating, modifying, manipulating, analyzing, and displaying signal information and attributes. Since the dawn of time, man has been the quintessential signal processor. Human signal processing was performed using one of the most powerful signal processing engines ever developed: the 25-W human brain that commits about 10 W to information processing. In that context, this biological processor is comparable to the Intel mobile Pentium III processor. As humans evolved, other agents were added to man's signal processing environment and repertoire, such as information coding in the form of intelligible speech, art, and the written word. In time, communication links expanded from local to global, global to galactic. It was, however, the introduction of electronics that enabled the modern information revolution. Analog electronics gave rise to such innovations as the plain old telephone system (POTS), radio, television, radar/sonar, and a host of other inventions that have revolutionized man's life and landscape. With the introduction of digital technologies over a half century ago, man has witnessed a true explosion of innovations that has facilitated the replacement of many existing analog solutions with their digital counterparts. In other instances, digital technology has enabled solutions that previously never existed. Included in this list are digital entertainment systems, digital cameras, digital mobile telephony, and other inventions. In some cases, digital technology has been a disruptive technology, giving rise to products that were impossible to envision prior to the introduction of digital technology. An example of this is the now ubiquitous personal digital computer.

ORIGINS OF DIGITAL SIGNAL PROCESSING (DSP)

Regardless of a signal's source, or the type of machine used to process that information, engineers and scientists have habitually attempted to reduce signals to a set of parameters that can be mathematically manipulated, combined, dissected, analyzed,

Digital Filters: Principles and Applications with MATLAB, First Edition. Fred J. Taylor.
© 2012 by the Institute of Electrical and Electronics Engineers, Inc.
Published 2012 by John Wiley & Sons, Inc.

1

or archived. This obsession has been fully realized with the advent of the digital computer. One of the consequences of this fusion of man and machine has been the development of a new field of study called *digital signal processing*, or DSP. Some scholars trace the origins of DSP to the invention of iterative computing algorithms discovered by the ancient mathematicians. One early example of a discrete data generator was provided in 1202 by the Italian mathematician Leonardo da Pisa (a.k.a. Fibonacci*). Fibonacci proposed a recursive formula for counting newborn rabbits, assuming that after mating an adult pair would produce another pair of rabbits. The predictive Fibonacci population formula is given by $F_n = F_{n-1} + F_{n-2}$ for the initial conditions $F_0 = 1$, $F_{-1} = 0$, and produces a discrete-time sequence that estimates the rabbit population $\{1, 1, 2, 3, 5, 8, 13, 21, 34, 55, \ldots\}$ as a function of discrete-time events. However, those who promote such action as evidence of DSP are overlooking the missing "D-word." DSP, at some level, must engage digital technology in a signal processing activity.

The foundations of DSP were laid, in fact, in the first half of the 20th century. Two agents of change were Claude Shannon and Harry Nyquist. They both formulated the now celebrated sampling theorem that described how a continuous-time signal can be represented by a set of sample values. Such representations were found to be so mathematically perfect that the original signal could be reconstructed from a set of sparsely distributed samples. Nyquist conjectured the sampling theorem in 1928, which was later mathematically demonstrated by Shannon in 1949. Their work

Claude Shannon (1916–2001)   **Harry Nyquist (1889–1976)**

* Fibonacci is short for *filius Bonacci*, son of Bonacci, whose family name means "good stupid fellow."

provided the motivation and framework to convert signals from a continuous-time domain to and from the discrete-time domain. The sampling theorem, while being critically important to the establishment of DSP, was actually developed prior to the general existence of digital technology and computing agents. Nevertheless, it was the sampling theorem that permanently fused together the analog and discrete-time sample domain, enabling what is now called DSP.

During the 1950s, and into the 1960s, digital computers first began to make their initial appearance on the technology scene. These early computing machines were considered to be far too costly and valuable to be used in the mundane role of signal analysis, or as a laboratory support tool by lowly engineers. In 1965, Cooley and Tukey introduced an algorithm that is now known as the fast Fourier transform (FFT) that changed this equation. The FFT was indeed a breakthrough in that it recognized both the strengths and weaknesses of the classic von Neumann general-purpose digital computer architecture of the day, and used this knowledge to craft an efficient code for computing Fourier transforms. The FFT was cleverly designed to distribute data efficiently within conventional memory architectures and perform computation in a sequential manner. Nevertheless, early adopters of the FFT would not necessarily have considered themselves to be DSP engineers since the field of DSP had yet to exist.

Since the introduction of the FFT, digital computing has witnessed a continuous growth, synergistically benefiting from the increasing computing power and decreasing cost of digital technologies in accordance with Moore's law.* The digital systems available in the 1970s, such as the general-purpose minicomputer, were capable of running programs that processed signals in an off-line manner. This process was often expensive, time-consuming, required considerable programming skills, and generally remained compute bound, limiting the type of applications that could be considered. During this epoch, early attempts witnessed the use of dedicated digital logic to build rudimentary digital filters and radar correlators for national defense purposes. These activities caused engineers and scientists to recognize, for the first time, the potential of DSP even though there was no formal field of study called DSP at that time. All this, however, was about to change.

In 1979, a true (albeit quiet) revolution began with the introduction of the first-generation DSP microprocessor (DSP μp) in the form of the Intel 2920, a device called an "analog signal processor" for marketing reasons. The 2920 contained on-chip analog-to-digital converter (ADC)/digital-to-analog converter (DAC), and a strengthened arithmetic unit that was able to execute any instruction in 200 μs. While initiating a fundamentally important chain of events that led to the modern DSP μp, by itself, the 2920 was a marketplace disappointment appearing in a few 300 b/s modems. It was, nevertheless, warmly embraced by a small but active group of digital audio experimenters. With the second generation of DSP μp (e.g., Texas Instruments TMS320C10), DSP technology exposed its potential value in a host of new applications. For the first time, products with embedded DSP capabilities became a practical reality establishing DSP as an enabling technology. The field,

---

* Moore's law $(Nt/A)(t_1) = (Nt/a)(t_0) \times 1.58^{(t_1 - t_0)}$ predicts that semiconductor density will double every 18 months.

now called DSP, rapidly developed in the form of academic programs, journals, and societies, and developing infrastructure technology. These beginnings swiftly gave way to a third and fourth generation of general-purpose DSP μp as well as custom DSP devices. Even though DSP remains a relatively young science, being only a few decades old, it has become both a major economic and technological force. DSP solutions are now routinely developed using commercial off-the-shelf (COTS) software and DSP μps and field programmable gate arrays (FPGAs), along with application-specific integrated circuits (ASICs).* There is now an abundance of DSP software design and development tools that serve this industry. Through the intelligent use of these resources, DSP has become an enabling technology for high-speed, low-cost data communications (modems), digital controllers, wireless solutions including cellular telephony and other personal communications services, video compression, multimedia and entertainment (audio and video solutions), plus a host of other applications. At the core of this revolution are the tens of thousands of scholars and technologists who now refer to themselves as DSP engineers and scientists. These engineers are hybrids in that they need to have competence in the application area they serve, to possess strong computer hardware and/or software skills, plus to have an understanding of the theory and practice of DSP. They, like DSP technology itself, are still in the formative stage. All that can be accurately predicted at this time is that DSP will be one of the principal technological driving forces of the 21st century economy.

SIGNAL DOMAINS

Signals are abundant in both the natural and artificial worlds. Nature, in particular, is rich in signals, from cosmic rays, bird trills, and the proverbial tree falling in the woods. Signals found in the natural world are produced by a variety of mechanisms. An example is a biological electrocardiogram (EKG) signal illustrated in Figure 1.1.

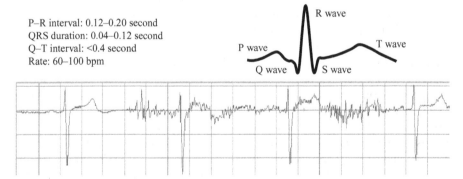

Figure 1.1   Typical EKG model (top) showing the P, Q, R, S, and T phases. Shown also is a typical EKG recording (bottom).

* ASICs are defined by the solution provider, not the manufacturer. A variation on this theme is the application-specific standard parts (ASSPs), which are essentially ASICs developed for high-volume commercial sale.

Figure 1.2   Two-dimensional image "Lena" as a $512 \times 512 \times 8$ bit/pixel image on the left and JPEG compressed image on the right.

Others include man-made signals such as those generated by musical instruments or the human voice. Artificial signals can be created without a natural production mechanism but using algorithms and electronic sound reproduction equipment. Signals from these domains can be classified as being one-, two-, and M-dimensional. A simple one-dimensional sinusoid $x(t) = \cos(\omega t)$ that can be expressed as an amplitude versus time trajectory. Images are often classified as being two-dimensional signals that can be expressed as a function of two spatial parameters, such as $f(x_1, x_2) = \cos(\omega x_1 + \omega x_2)$. A simple black and white image can be represented as an array of image values $i(x, y)$, where $i(x, y)$ is the image intensity at coordinates (x, y) (see Fig. 1.2). Signals of higher dimension also exist. The Dow Jones industrial average, for example, is a function of 30 economic variables and represents a multidimensional signal.

Causality is also an important signal property. Causal signals are produced by causal systems (nonanticipative systems) where the output signal output $y(t)$, at some specific instant t_0, depends only on the system input $x(t)$ for $t \leq t_0$. Signals that are not causal are called noncausal, anticausal signals, or anticipatory.* While noncausal signals are not products of a physically realizable signal generation mechanism, they will play a significant role in the mathematical study of signals and systems or in performing off-line simulations. For example, the signal $x(t) = \cos(\omega_0 t)$ technically persists for all time (i.e., $t \in (-\infty, \infty)$) and is therefore noncausal. It is also recognized that $x(t) = \cos(\omega_0 t)$ represents a mathematically important signal even though it could not have been created by a physical signal generator.

SIGNAL TAXONOMY

Any meaningful signal taxonomy needs to recognize that signals can live in different domains. Specifically, there are three important signal domains, and they are developed below.

---

* He who answers before listening—that is his folly and his shame—Proverbs 18:13.

Continuous-Time Signals

Continuous-time or analog signals $x(t)$ are defined as a continuum of points in both independent and dependent variables. Both the signal's amplitude $x(t)$ and time instance t are numbers known with infinite precision. Continuous-time signals can be further portioned into differentiable, analytic, piecewise differentiable, continuous, piecewise constant, as well as others.

Discrete-Time Signals

A discrete-time signal $x[k]$ is obtained by sampling a continuous-time signal $x(t)$ with an ideal sampler. Specifically, if T_s denotes the sample period, then $f_s = 1/T_s$ is called the sample rate or sample frequency and is measured in sample per second, denoted "Sa/s," although "Hz" is commonly used interchangeably with Sa/s. The sample value at the sample instance $t = kT_s$ is denoted $x(t = kT_s) = x[k]$. A collection of such sample values is called a time series. A discrete-time series consists of sample values that are continuously resolved along the dependent axis (amplitude) and discretely resolved along the independent axis (time). Whereas the sample instances are discrete (in time), the sample value $x[k]$ is an infinite precision real or complex number. Discrete-time signals can be physically created by passing a continuous signal through an electronic device called an impulse sampler or ideal sampler. The sampled value $x[k]$ can be used to construct a continuous-time signal $y(t)$ using an inverse process called interpolation. An example interpolator is a called sample-and-hold (S/H) circuit, as shown in Figure 1.3. S/H circuits are commonly found in the design of DACs. Other forms of interpolation are possible.

Computing algorithms, such as those studied in discrete mathematics, can be used to produce discrete-time signals. In addition, discrete-time series also arise in the fields of economics, biology, calculus, statistics, physics, plus others. The engineering importance of discrete-time signals can be traced back to a post-World War II era in the form of sampled data control systems and telephony. During the early days of the Cold War, strategic bombers were flying missions having long time durations while attempting to navigate with a high degree of accuracy. This was a challenging problem for the day's analog control systems. Small drifts in the control surface signal values could accumulate over time, resulting in large positional errors. What was required was a more precise autopilot technology. It was discovered that if the control signal was sampled and modulated, drift-free alternating current (AC) amplifiers could replace the troublesome drift prone direct current (DC) amplifiers. The simplest modulation scheme is an alternating sign periodic analog pulse train using a device that was called a "chopper." The modulated signal was then transmit-

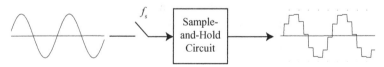

Figure 1.3   Discrete-time signal sampler and sample-and-hold circuit.

ted to a receiver where it was demodulated, returning all the samples to their original sign. This gave rise to a technology called sampled-data control, which had a strong following in the 1950s and 1960s.

The most enduring technology emerging from this era is found in telephony. It was discovered that a number of distinct discrete-time time series could be inter-laced (i.e., time-division multiplexed) onto a common channel, thereby increasing the channel's capacity in terms of the number of subscribers per line per unit time. The result was that the telephone company could bill multiple clients for using a single copper line. Claude Shannon developed the mathematical framework by which these signals can be time multiplexed, transmitted along a common line, and reconstructed at each individual receiver. Shannon's innovation, known as Shannon's sampling theorem, has been a driving force behind most DSP techniques and methodologies.

Digital Signals

Digital signals are discrete-time signals that are also quantized along the dependent axis (amplitude). Digital signals can be produced by a digital computer using finite precision arithmetic or by passing an analog signal $x(t)$ through an ADC or A/D, also producing a finite precision approximation of a discrete-time signal. In either case, quantizing the amplitude of the original signal introduces an uncertainty called quantization error. Controlling and managing such errors is often critical to the successful design of a DSP solution.

A general signal taxonomy is presented in Figure 1.4. Contemporary signal processing systems typically contain a mix of analog, discrete, and digital signals and systems. A signal's original point of origin is often the continuous-time or analog domain. Digital signals, however, are becoming increasingly dominant in this mix. Applications that were once considered to be exclusively analog, such as sound

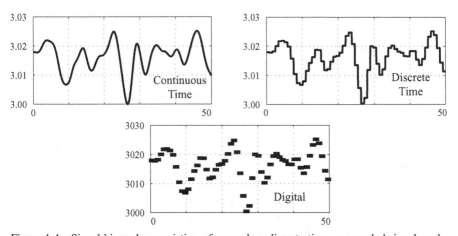

Figure 1.4   Signal hierarchy consisting of an analog, discrete-time or sampled signal, and digital or quantized signal process.

recording and reproduction, have become digital. The wireless communications industry are replacing analog components in radios, as well as back-end audio and signal decoding sections, with digital devices. Images and video signals are now routinely coded and decoded as digital signals. Discrete-time systems, as defined, are rarely found in use today except as part of the sampling subsystems (samplers) found in ADCs. The reason for this paradigm shift from analog signal processing to DSP is primarily due to two breakthroughs. The first is the sampling theorem and the second is the product of the fruitful digital semiconductor industry. Once this bridge was crossed, it became logical to replace everything possible with digital technology.

DSP: A DISCIPLINE

Signal processing is a gift of the sampling theorem and formidable armada of companion theories, methodologies, and tools, such as the celebrated FFT. Initially DSP was envisioned simply as an analog replacement technology. It is now clearly apparent to many that DSP will move into new areas and become the dominant signal processing technology in the 21st century. DSP has matured to the point where it can claim to be an academic discipline replete with a rich infrastructure industry. DSP continues to gain semiconductor market share mainly because it can deliver solutions. The DSP advantage is summarized below.

Digital Advantages

The attributes of a digital solution are as follows:

- Both analog and digital systems can generally be fabricated as highly integrated semiconductor systems. Compared with analog circuitry, digital devices can take full advantage of submicron technologies and are generally more electronically dense, resulting in both economic and performance advantages.

- As semiconductor technologies shrink (deep submicron) and signal voltages continue to decline (1.25 V and lower), the intrinsic signal-to-noise ratio found at the transistor level decreases. Digital systems are far more tolerant of such internal noise. These devices, however, are essentially useless as an analog system (e.g., equivalent 3-bit precision per transistor).

- Digital systems can operate at extremely low frequencies, which would require unrealistically large capacitor and resistor values if implemented as an analog solution.

- Digital systems can be designed with increased precision with only an incremental increase in cost, whereas the precision of an analog system precision is physically limited (10 bits ~ 60-dB dynamic range typical).

- Digital systems can be easily programmed to change their function whereas reprogramming analog systems is extremely difficult.

- Digital signals can be easily delayed and/or compressed, an effect which is difficult to achieve with analog signals.

- Digital systems require no external alignment, while analog systems need periodic adjustment (due to temperature drift, aging, etc.).

- Digital systems do not have impedance-matching requirements, while analog systems do.

- Digital systems, compared with analog devices, are less sensitive to additive noise as a general rule.

Analog Advantages

There are, however, a few analog attributes that resist the digital challenge. They are as follows:

- Analog systems can operate at extremely high frequencies (e.g., microwave and optical frequencies) that exceed the maximum clock rate of a digital device or ADC.

- Analog solutions are sometimes more cost effective (e.g., first-order resistor-capacitor [RC] filter) compared with solutions fashioned with digital components (e.g., ADC, digital filter, plus DAC).

Driven by advancements in semiconductors, software, and algorithms, DSP will be a principal enabling and facilitating technology in the following areas:

Audio

Audio-video receivers

Computing

Digital radio

Home audio

Flat panel displays

Internet audio

Pro audio

Speech

Toys

Transportation

Chassis sensors

Power train

Driver displays

Security systems

Safety systems

Broadband

Wireless local area network (LAN)

Cable

Digital subscriber line (DSL)

Voice-over Internet protocol (VoIP)

Control

Digital power supplies

Embedded sensors

Industrial drives

Motors

Instrumentation

Medical

Automated external defibrillators

Monitoring

Hearing aids

Imaging

Prosthetics

Military

Avionics

Countermeasures

Imaging

Munitions

Navigation

Radar/sonar

Wireless

Handsets

Infrastructure

Radiofrequency tagging

Security

Biometrics

Smart sensors

Telecom

High-frequency radios

Infrastructure

Navigation	Digital TV	Set-top boxes
Telecom accessories	Digital video	Streaming media
Wire systems	Digital recorders	Surveillance
Video and Imaging	Internet protocol (IP) video phones	Video conferencing
Still cameras	Media devices	Vision

Infrastructure

DSP engineers and technologists continue to invent new products based on preexisting and emerging DSP theory and technology. Based on what has been witnessed over the brief history of DSP, one can only view in awe the possibilities of the future.

CHAPTER 2

SAMPLING THEOREM

INTRODUCTION

One of the most important scientific advancements of the first half of the 20th century is attributable to Claude Shannon of Bell Laboratories. Many of Shannon's inventions remain with us today. One of his more amusing creations was a black box that, when activated with a switch, would extend a green hand outward and turn the switch off. Of greater value is his celebrated and enduring sampling theorem. Shannon's interest in sampling can be attributed to the fact that he worked for the telephone company. He was therefore interested in maximizing the number of billable subscribers that could simultaneously use a copper telephone line, the technology of the day. Shannon's innovation was to sample the individual subscriber's conversations, interlace the samples with samples from other subscribers, place them all on a common telephone wire, and finally reconstruct the original message at the receiver after de-interlacing the samples. Today, we refer to this process as time-division multiplexing (TDM). Shannon established the rules that govern the sampling and signal reconstruction procedure. Without a reconstruction rule, however, Shannon's labors would have held no value to the telephone company. The outcome was the sampling theorem, which has become the core to understand the theory and practice of digital signal processing (DSP). The theorem both enables and constrains the performance of the typical DSP system suggested in Figure 2.1. The diagrammed system consists of an analog-to-digital converter (ADC), digital-to-analog converter (DAC), digital or DSP processor, plus analog signal conditioning filters (i.e., anti-aliasing and interpolation filter). The sampling theorem also motivates the need for these signal conditioning filters.

SHANNON'S SAMPLING THEOREM (AN ENABLING TECHNOLOGY)

The sampling theorem states that if a band-limited signal $x(t)$, whose highest frequency component is bounded from above by some f_{max}, is periodically sampled at some rate f_s, where

Digital Filters: Principles and Applications with MATLAB, First Edition. Fred J. Taylor.
© 2012 by the Institute of Electrical and Electronics Engineers, Inc.
Published 2012 by John Wiley & Sons, Inc.

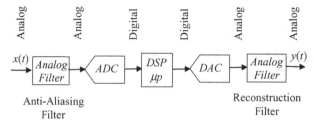

Figure 2.1   DSP system consisting of an input signal conditioner (anti-aliasing filter), ADC, DSP microprocessor, DAC, and output signal conditioner (interpolation filter).

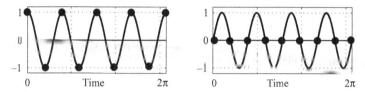

Figure 2.2   Sample rate hypothetical. Cosine sampled at f_s, (left) and sine sampled at f_s (right).

$$f_s > 2 \times f_{max}, \tag{2.1}$$

then the original signal $x(t)$ can be reconstructed from the sample values $x(t = kT_s) = x[k]$, where T_s (sample period) satisfies $T_s = 1/f_s$. The parameters that define this process are $2 \times f_{max}$, called the Nyquist sample rate in sample per second (Sa/s), and $f_s/2$, referred to as the Nyquist frequency in hertz. It should be strongly noted that the sampling rate must be greater than the Nyquist sample rate $(2f_{max})$ and not equal to $2f_{max}$. The importance of the statement can be illustrated by considering a simple cosine wave $x(t) = \cos(\pi f_s t)$. If sampled at a rate f_s, beginning at $t = 0$, the result is the time series $x[k] = \cos(\pi k) = \{1, -1, 1, \ldots, (-1)^k, \ldots\}$ as illustrated in Figure 2.2. It is tempting to assume that a continuous-time signal with a cosine envelope could have been reconstructed from the displayed sample values. Is this then a counterexample of Shannon's sampling theorem? Consider a slight modification of the previous observation by using $x(t) = \sin(\pi f_s t)$. Sampling $x(t)$ at the same rate f_s, starting at $t = 0$ would produce a time series $x[k] = \sin(\pi k) = \{0, 0, 0, \ldots\}$, which would be reconstructed (incorrectly) as $x(t) = 0$. It should be understood exactly what Shannon said and what he did not say. Shannon's sampling theorem simply states that if you sample at a rate $f_s > 2 \times f_{max}$, then the original signal (parent) can be theoretically reconstructed from its sample values $x[k]$. It makes no claims beyond this. What remains to be established is a means of reconstructing the original signal from its sample values. This process is called interpolation.

SIGNAL RECONSTRUCTION

Shannon assumed that an analog signal $x(t)$ can be replaced with a set of periodic sample values $x[k]$ that form what is called a time series. The reconstruction of $x(t)$

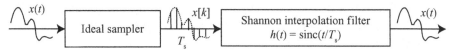

Figure 2.3 Shannon's interpolation process showing how a signal's sample values are converted (interpolated) into a continuous-time signal using an ideal sinc filter.

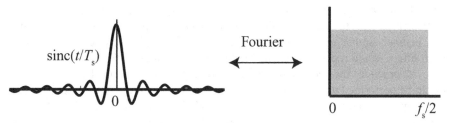

Figure 2.4 Shannon's interpolating filter interpreted in the continuous-time (left) and continuous-frequency domains (right).

from its sample values involves the use of an interpolating filter. According to Shannon, the ideal continuous-time interpolation filter is

$$h(t) = \sin\left(\frac{\pi t}{T_s}\right) \bigg/ \frac{\pi t}{T_s} \triangleq \mathrm{sinc}\left(\frac{t}{T_s}\right). \tag{2.2}$$

Formally, the output of a Shannon interpolator to an arbitrary input time series $x[k]$ is mathematically defined by a process called linear convolution as motivated below:

$$x(t) = \sum_{k=-\infty}^{\infty} x[k]h(t - kT_s) = \sum_{k=-\infty}^{\infty} x[k]\frac{\sin(\pi(t - kT_s)/T_s)}{(\pi(t - kT_s)/T_s)}. \tag{2.3}$$

The interpolation process, described in Equation 2.3, is graphically interpreted in Figure 2.3. The input signal $x(t)$ is sampled at a rate consistent with Shannon's sampling theorem to produce a time series $x[k]$. The time series is then passed through Shannon's interpolation filter that performs the convolution operation described by Equation 2.3. The outcome is the reconstructed signal $x(t)$, which again resides in continuous-time domain.

Shannon's sinc interpolating filter, defined by Equation 2.2, can also be interpreted in the frequency domain as suggested in Figure 2.4. The frequency domain envelope of Shannon's interpolating filter is that of an ideal low-pass (boxcar) filter having a passband defined over $f \in [0, f_s/2)$ Hz (i.e., 0 Hz [DC] to the Nyquist frequency).

SHANNON INTERPOLATION

Assume that an analog sinusoidal signal $x(t) = \sin(2\pi f_0 t)$ is sampled above the Nyquist sample rate at a frequency f_s, where $f_s > 2f_0$. The sampling process produces a discrete-time signal $x[k]$. Theoretically, a perfect reconstruction of an analog signal $x(t)$ from that signal's sample values requires that Shannon's interpolating sinc filter be employed as shown in Figure 2.5.

While Shannon's interpolating filter is elegant, it presents a major implementation obstacle. The interpolating filter $h(t) = \mathrm{sinc}(t/T_s)$ is seen to be noncausal by virtue of the fact that the filter's response exists for all time, $-\infty \le t \le \infty$. As a result, Shannon's interpolation filter is not physically realizable for real-time application since the filter's impulse response exists in prehistory (i.e., $t \le 0$). This has caused DSP technologists to search for alternative real-time interpolating filters that behave in a manner similar to a Shannon interpolator, but are also physically realizable. Such filters would replace Shannon's filter shown in Figure 2.5.

In practice, the ideal signal interpolation process, described in Figure 2.5, would be performed using the technology as shown in Figure 2.6. It is required that the sampled signal be first converted into a digital data stream using an ADC. The digitized samples are converted back into analog form using a DAC, which translates digital words into analog signal levels that are maintained for a full sample period (sample and hold [S&H]). This process is called a zero-order hold (ZOH). The

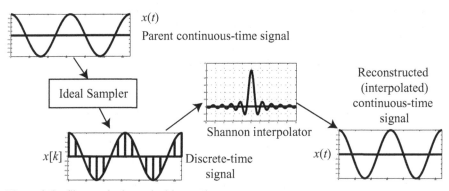

Figure 2.5   Shannon's theoretical interpolator process.

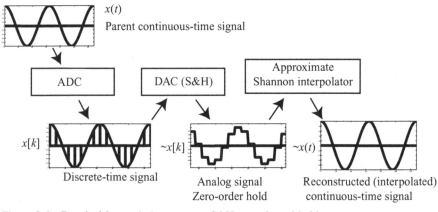

Figure 2.6   Practical interpolation process. S&H, sample and hold.

analog ZOH signal is then presented to a causal analog filter that approximates the behavior of a Shannon interpolator. The output is an interpolated analog signal that approximates the original input $x(t)$. Interpolation errors are due to quantization effects introduced by ADC and DAC devices, plus the differences between the Shannon and approximate interpolation filters.

Ultimately, the objective of the interpolating filter is to produce an output that is in good agreement with the analog signal of origin, given knowledge only of the analog signal's sample values. Traditionally, this has been accomplished, to varying degrees of success, using one or more of the following strategies:

- a ZOH circuit produces an output $y(t)$ that is a piecewise constant approximation of the input sample value $x[k]$, where $y(t) = x[k]$ for $t \in [kT_s, (k + 1)T_s]$;
- a first-order hold circuit produces an output $y(t)$ that is a piecewise linear approximation of the original signal passing through $x[k]$ and $x[k + 1]$, where $y(t) = x[k] + t(x[k+1] - x[k])/T_s$, for $t \in [kT_s, (k + 1)T_s]$;
- higher-order interpolation schemes (polynomials, splines, etc.);
- a low-pass analog "smoothing" filter.

The anticipated attributes of the classic analog "smoothing" filter are interpreted in Figure 2.7 in both time and frequency domains. The frequency domain response of the physically realizable filter, having an impulse response $h(t)$, is shown in Figure 2.7, and can replace the ideal noncausal Shannon interpolator. In practice, the physically realizable filter would exhibit noticeable roll-off in transition band. From a practical design standpoint, typical low-pass interpolating filters have passbands extending out to $0.4f_s$ Hz. Beyond that frequency, the analog filter transitions into the stopband whose depth is a function of filter order. If the signal being interpolated is dominated by low-frequency signal components, then the filter's passband can be made smaller than $0.4f_s$ Hz (i.e., aggressive low-pass filtering). For the case where the sample frequency is fixed, realizable active filters are normally employed that are based on the use of common operational amplifiers (Op Amps). These filters are well documented in the literature and are supported with numerous software design packages. An example is the fixed-frequency Sallen–Key interpolating filter.

The Sallen–Key filter is commonly used in a fixed-frequency interpolation role. A second-order Sallen–Key filter is shown in Figure 2.8 and belongs to a class of active devices called voltage-controlled voltage source (VCVS) filters. VCVS filters use a unity-gain amplifier having effectively infinite input impedance and zero

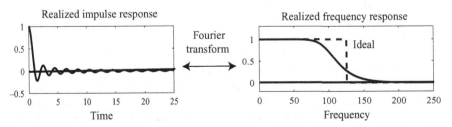

Figure 2.7   Physically realizable interpolating filter.

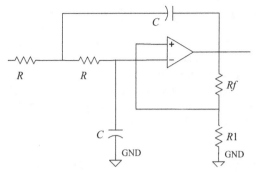

Figure 2.8   Active Sallen–Key filter. GND, ground.

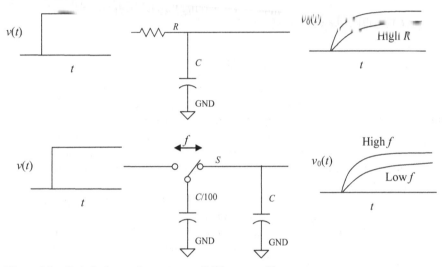

Figure 2.9   Switched capacitance versus RC low-pass filter.

output impedance. Filter design parameters are tabled or obtained from a predefined computer program. The choice of parameters determines whether a Bessel, Butterworth, or Chebyshev filter type is to be realized.

Fixed coefficient interpolation filters are often too rigid to serve in multisample rate environments, such as those found in many audio and multimedia applications. Developing programmable sample rate filters using conventional electronic Op-Amp-enabled filters is expensive and technically challenging. Instead, programmable analog filters are often fabricated using a switched capacitance technology. A switched capacitance filter connects two capacitors of different values (typically 1:100) through an electronic switch. The filter performance is determined by a charge transfer process and the switching rate. The response of a switch capacitance filter emulates that of a low-pass RC filter as suggested in Figure 2.9. The realized programmable filter therefore has an RC low-pass response that is needed to perform low-pass interpolation.

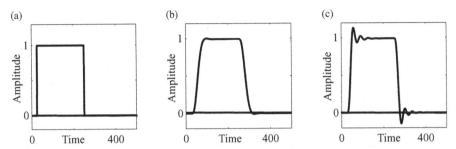

Figure 2.10 Pulse response of a Bessel and Chebyshev low-pass filter. (a) Pulse waveform; (b) Bessel filter; (c) Chebyshev filter.

Low-pass interpolation electronic (e.g., Op Amp) or switched capacitance filters can be used to design low-pass Bessel, Butterworth, Chebyshev, or Elliptic analog interpolating filters. Figure 2.10 exhibits the response of a typical Bessel and Chebyshev filter model to a piecewise constant pulse input (i.e., that of a 0th order hold). The Bessel filter (similar to a Butterworth filter) has a tendency to be slow to respond to rising and trailing signal edges but is devoid of any overshoot such as that associated with a Chebyshev and elliptic filter.

SAMPLING MODALITIES

In practice, ADC sampling occurs in one of the following three modes:

- critically sampled: $f_s = 2f_{max}$,
- oversampled: $f_s > 2f_{max}$,
- undersampled: $f_s < 2f_{max}$,

where it is assumed that the highest frequency component in an input signal $x(t)$ is bounded from above by f_{max}. Critical sampling presumes that sample rate is set to $f_s = 2f_{max}$, which technically does not satisfy the sampling theorem. To insure that the analog signal's highest frequency is essentially bounded by f_{max}, an analog pre-filter, called an anti-aliasing filter, is often placed between the signal and ADC as shown in Figure 2.1. A successful anti-aliasing filter is expected to significantly suppress signal energy residing beyond the Nyquist frequency $f_s/2$ Hz. However, it is extremely difficult and expensive to construct a "steep-skirt" or narrow transition band analog filter that can achieve high frequency selectivity. Using a commonly available 44.1 kSa/s ADC to process a 20-kHz audio record, shown in Figure 2.11, results in a 22.05-kHz Nyquist frequency and a 2.05-kHz transition band. An eighth-order Butterworth filter having a −3-dB passband gain at 20 kHz could only manage a −5 dB gain at 22.05 kHz. The system designer may therefore have to consider an alternative strategy. Oversampling, for example, relaxes the design requirements imposed on the analog anti-aliasing filter. This allows the Nyquist frequency to be set well into the analog filter's stopband, resulting in a practical anti-aliasing filter. Using a 4× oversampling solution, for example, would set the sample rate to 160 kHz,

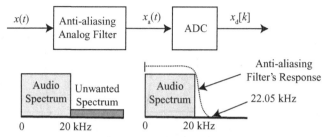

Figure 2.11   Hypothetical multimedia data acquisition system.

producing an 80-kHz Nyquist frequency. The anti-aliasing filter's transition band now resides between 20 k and 80 kHz, or is 60 kHz wide instead of 2.05 kHz for the 11.1 kSa/s case. An eighth-order Butterworth anti-aliasing filter can achieve a –48 dB gain at the end of the transition band. In the case of recorded speech or music, it is assumed that any signal residing between 20 k and 80 kHz is already of low amplitude, well below the passband signal levels. The result would be negligible signal energy found at the output of the anti-aliasing filter beyond the Nyquist frequency.

Example: Sampling

The system characterized in Figure 2.11 would suggest that 44.1 kSa/s is probably too low a sample rate for use in high-end audio signal processing. The question arises regarding the rationale behind standardizing on 44.1 kSa/s for multimedia applications. The answer is that low-cost 44.1 kSa/s ADCs preexisted the multimedia era and were in plentiful supply. These pre-multimedia ADCs were extensively used to support video recording industry standards, namely,

- NTSC—490 lines/frame, 3 samples/line, 30 frames/s = 44,100 Sa/s;
- PAL—588 lines/frame, 3 samples/line, 25 frames/s = 44,100 Sa/s.

The engineering choice of the 44.1 kSa/s standard, therefore, was based on economics.

The maximum sample speed is set by technology. Signals can now be digitized far in excess of 1 GSa/s. In general, ADC speed is gained at a cost in power dissipation and high cost. At the other end of the spectrum are low sample rate applications. A biomedical signal, for example, may be band-limited to a few hertz. Building an ADC that can operate at such slow sample speeds is extremely challenging and expensive (e.g., 8 Sa/s). Instead, the designer would be wise to choose to use a low-cost ADC, which would oversample the signal by a factor of 1000 (e.g., 8 kSa/s), and then keep only 1 in 1000 samples. Here, the fact that a signal is highly oversampled is justified in order to produce a low-cost, low-complexity solution. Finally, undersampling can be intentional or unintentional. In either case, the sampling theorem is violated. In such instances, a phenomenon called aliasing can and will occur. Aliasing, developed in Chapter 3, can introduce serious and disturbing errors into a DSP solution. While the introduction of aliasing errors can degrade system performance, intelligently controlled aliasing can also be used as a design asset.

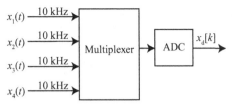

Figure 2.12   Multichannel ADC.

MULTICHANNEL SAMPLING

In certain circumstances, multichannel signals obtained from a multichannel analog multiplexer can be converted using a single ADC operating at an oversampled rate. This action can be motivated in the context of the application shown in Figure 2.12, which requires that four independent signals be digitized with 16 bits of accuracy. The baseband bandwidths for the four signals are 10 k, 10 k, 10 k, and 30 kHz, respectively. Suppose the ADC is capable of multiplexing between the channels. If the ADC runs at 120 kSa/s, then the ADC period is $T_s = 1/f_s = 8.33$ μs. In order to sample the signal with the single multiplexed ADC, the samples will need to be interlaced as $\{\ldots, x_1(0T_s), x_4(1T_s), x_2(2T_s), x_4(3T_s), x_3(4T_s), x_4(5T_s), x_1(6T_s), \ldots\}$. It can be noted that the signal $x_4(t)$ is being polled at a 60-kHz rate and the other three at 20-kHz rate each.

MATLAB AUDIO OPTIONS

MATLAB provides users with a limited ability to play recorded audio records through a standard multimedia soundboard. Information regarding the use of these tools can be displayed using the MATLAB help command:

» help sound

SOUND(Y,FS) sends the signal Y (with sample rate FS) to multimedia speakers. The array Y is restricted to the dynamic range ±1.0. Values outside that range are clipped. Stereo sounds can be played on platforms that support stereo, where Y is an Nx2 matrix. SOUND(Y) plays the audio record at a default sample rate of 8192 Hz, and SOUND(Y,FS,BITS) assigns BITS bits/sample where most platforms support BITS = 8 or 16 processing.
    A similar family of functions can be viewed using

» help soundsc

SOUNDSC—autoscale and play vector as sound,
    SOUNDSC—performs an autoscaling of the data array, playing the audio record as loud as possible without clipping.

CHAPTER *3*

ALIASING

INTRODUCTION

Whenever Shannon's sampling theorem is violated, a phenomenon called aliasing can occur. Aliasing manifests itself as a corruption of a reconstructed (interpolated) signal or image. Specifically, aliasing occurs when a reconstructed signal or image impersonates another signal or image. To illustrate, consider as a youth seeing your first Hollywood western. The camera was following a moving stagecoach at 30 frames/s rate. The stagecoach wheel was moving at a leisurely clockwise rate of 45° per frame, and the viewer sees the wheel spinning at a constant rate of (30/8) revolutions per second in a clockwise direction as shown in Figure 3.1. Then danger appears and the background music is translated to a minor key. The stagecoach is now in a headlong rush moving forward at full speed; the wheel is spinning at a rate of 315° per frame. From the viewer's perspective, it would appear that the wheel is turning at a rate of −45° (315° − 360°) per fame in a counterclockwise direction as suggested in Figure 3.1. That is, the fast-moving wheel now impersonates a slow-moving wheel rotating in the opposite direction. The impersonation effect is called aliasing.

ALIASING

Shannon's sampling theorem states that a signal can be reconstructed (interpolated) from its sample values without error provided that it is sampled above the Nyquist sample rate $2 \times f_{max}$. If this condition is violated by choosing too small a sample rate, then signal reconstruction is compromised. To illustrate, consider the monotone cosine wave shown in Figure 3.2. The cosine is seen to be sampled well below the critical sampling rate (i.e., undersampled) and results in a reconstructed analog signal that continues to be sinusoidal but at a much lower frequency. This observation suggests that an important metamorphosis has taken place, one that maps a sinusoidal signal originally located at a frequency ω_0 to a lower baseband frequency $\Delta_0 \in (-f_s/2, f_s/2)$. What is desired is to develop a mathematical framework that can predict the baseband frequency of a reconstructed (interpolated) signal if the sample rate and original signal frequency are known.

Digital Filters: Principles and Applications with MATLAB, First Edition. Fred J. Taylor.
© 2012 by the Institute of Electrical and Electronics Engineers, Inc.
Published 2012 by John Wiley & Sons, Inc.

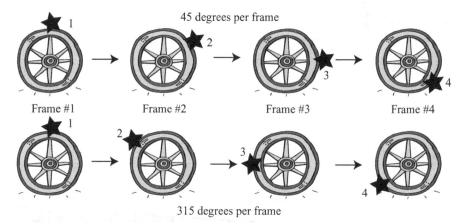

45 degrees per frame

Frame #1   Frame #2   Frame #3   Frame #4

315 degrees per frame

Figure 3.1   Image aliasing experiment showing a slowly moving stagecoach (top) and rapidly moving stagecoach (bottom).

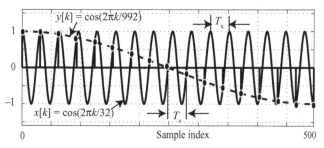

Figure 3.2   Original sine wave with a period of 32 seconds being sampled at a rate 1/31 Sa/s. The sample values outline a sinusoid having a period of 992 samples.

The effects of aliasing can be experimentally motivated by sampling a frequency selectable sinusoid $x(t) = \cos(2\pi f_0 t)$ at a fixed rate f_s. For the case where the sinusoid's frequency is set to $f_0 = \alpha f_s$, the sampled and reconstructed signals are shown in Figure 3.3 for specific values of $\alpha = 0, 0.1, 0.3$, and 0.8 corresponding to f_0 equal to 0%, 10%, 30%, and 80% of the sampling frequency f_s. It should be recalled that Shannon's sampling theorem is satisfied whenever f_0 is less than 50% of f_s ($\alpha < 0.5$). In those cases, the original signal can be seen to be recovered (using the Shannon interpolation) from the original signal's sample values. However, when the input signal is at a frequency equal to 80% of the sample frequency, the original signal is no longer evident as shown in Figure 3.3 for $\alpha = 0.8$. In fact, a much lower-frequency sinusoid appears as the interpolated signal. That is, the sample values of the high-frequency signal ($\alpha = 0.8$) appear to be impersonating the sample values of a much lower baseband signal. The objective is to determine the new baseband frequency.

It had been established that aliasing refers to a condition in which the sample values of one signal impersonates another signal. At issue is the form the alias signal takes when aliasing occurs. The form can be determined with a simple analytical

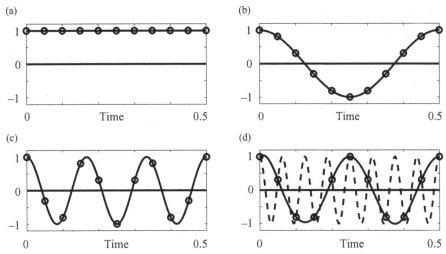

Figure 3.3   Signal sampling experiments where the input signal frequency is 0%, 10%, 30%, and 80% of the sampling frequency f_s. The individual sample values of the 80% case are seen to impersonate the sample values of a much lower frequency sinusoid. (a) Analog frequency = 0.0 (i.e., DC); (b) analog frequency = 10% of f_s; (c) analog frequency = 30% of f_s; (d) analog frequency = 80% of f_s.

study using a sinusoid sampled above and below the Nyquist sample rate. First, consider oversampling a simple monotone sinusoidal signal operating at a frequency $f_0 = 0.3f_s(\alpha = 0.3)$, thereby satisfying the sampling theorem. Upon sampling $x(t) = \cos(2\pi f_0 t)$ at a rate f_s, one obtains the time series $x[k] = \cos(0.6\pi k)$. The resulting reconstructed or interpolated continuous-time signal, based on the discrete-time samples shown in Figure 3.3 ($\alpha = 0.3$), is seen to be a faithful replica of the original signal $x(t)$. Now consider increasing the signal's frequency to $f_1 = 0.8f_s(\alpha = 0.8)$. In this case, Shannon's sampling theorem is violated, and the signal is said to be undersampled. A Shannon interpolator, however, would be unaware of this heresy and assumes that the input signal has been properly sampled. Upon sampling $x(t) = \cos(2\pi f_1 t)$, the time series $x[k] = \cos(2\pi f_1 kT_s) = \cos(1.6\pi k)$ results. The outcome of this process is displayed in Figure 3.3 ($\alpha = 0.8$), which shows that the reconstruction of a sampled signal to be at a much lower frequency. The new frequency can be determined using the trigonometric identity $\cos(a + b) = \cos(a)\cos(b) - \sin(a)\sin(b)$, where $a + b = 1.6$, $a = 2$, and $b = -0.4$, then $x[k] = \cos(1.6\pi k) = \cos((2 - 0.4)\pi k) = \cos((2)\pi k) \ \cos((-0.4)\pi k) - \sin((2)\pi k) \ \sin((-0.4)\pi k) = \cos((-0.4)\pi k)$. The claim is that the resulting baseband frequency is actually $-0.2f_s(\alpha = -0.2)$. Since cosine is an even function, it can also be claimed for $\cos((-0.4)\pi k) = \cos((0.4)\pi k)$, and that the reconstructed signal has a frequency of $0.2f_s$. Repeating the analysis for $f_1 = 0.8f_s(\alpha = 0.8)$ and $x(t) = \sin(2\pi f_1 t)$, and using the identity $\sin(a + b) = \sin(a)\cos(b) + \cos(a)\sin(b)$, it follows that $x[k] = \sin((-0.4)\pi k)$, which corresponds to a baseband frequency $f = -0.2f_s(\alpha = -0.2)$. Since sine is an odd function, it can also be claimed for $\sin((-0.4)\pi k) = -\sin((0.4)\pi k)$, introducing a 180° phase shift in the reconstructed signal.

Modular Arithmetic

The previous narrative develops the thesis that undersampling results in an aliased reconstructed signal. The reconstructed signal, using an ideal Shannon interpolator, is a signal whose frequencies reside exclusively within the baseband range $f \in (-f_s/2, f_s/2)$. The actual baseband frequency of the aliased signal can be determined using a modular arithmetic equation*:

$$f_1 = f_0 \bmod(f_s), \; f_1 \in [-f_s/2, f_s/2), \tag{3.1}$$

where

f_0 = frequency of the sampled signal,

f_s = sample rate,

f_1 = baseband frequency of the interpolated sampled signal.

Based on this analysis, if the discrete-time analysis of sinusoidal signals $x(t) = \cos(\omega_0 t)$ and $x(t) = \sin(\omega_0 t)$ are sampled at a rate f_s, then the baseband outcomes are shown in Table 3.1. To illustrate its use in a problem environment, consider an electrocardiogram (EKG) signal sampled at a rate of 135 Sa/s signal as shown in Figure 3.4. The biologically significant signal information is concentrated about 0 Hz. However,

TABLE 3.1.   Aliasing Study (Sample Rate f_s Sa/s)

f_0 (Input Frequency)	$f_1 \in [-f_s/2, f_s/2)$ (Interpolated Frequency)	Aliased
$0.2f_s$	$0.2f_s \bmod(f_s) = 0.2f_s$	No
$0.4f_s$	$0.4f_s \bmod(f_s) = 0.4f_s$	No
$0.6f_s$	$0.6f_s \bmod(f_s) = -0.4f_s$	Yes
$0.8f_s$	$0.8f_s \bmod(f_s) = -0.2f_s$	Yes
$1.0f_s$	$1.0f_s \bmod(f_s) = 0.0f_s$	Yes
$1.2f_s$	$1.2f_s \bmod(f_s) = 0.2f_s$	Yes
$1.4f_s$	$1.4f_s \bmod(f_s) = 0.4f_s$	Yes

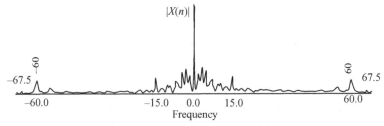

Figure 3.4   EKG spectrum showing the effects of aliasing. Unexpected spectral lines are shown at ±15 and ±60 Hz.

---

* $A = B \bmod(C)$ if $B = kC + A$, k an integer. For example, $9 = 109 \bmod(10)$.

there is strong evidence of spectral contamination of some type at several higher-frequency locations. The instrument producing the spectrum calibrates the baseband frequency axis to range over $f \in [-f_s/2, f_s/2) = [-67.6, 67.5)$ as shown. Two suspicious positive baseband tones are located at 60 and 15 Hz. The 60-Hz tone can be attributed to 60-Hz line-frequency interference, probably due to poor ground isolation. The 15-Hz contamination remains a mystery until aliasing is suspected. Specifically, the possible sources of a 15-Hz aliased signal sampled at 135 Hz are signals having frequencies $\{\ldots, -255, -120, 15, 150, 285, \ldots\}$. All of these frequencies satisfy the modular equation $15 = f_0 \bmod(135)$. Of these, the -120-Hz line stands out as being physically significant. A similar search suggests the presence of a $+120$-Hz signal being present as well as shown in Figure 3.4.

CIRCLE CRITERIA

Care needs to be taken in the production of modular arithmetic operations using generic tools such as MATLAB. To illustrate, for $f_s = 1$ kHz, and $f_0 = 750$ Hz, then f_1, is a baseband frequency residing between $f_1 \in [-500, 500)$, having a value of -250 Hz. However, MATLAB's interpretation is 750mod(1000)=750, which is not a valid baseband frequency. This problem can be mitigated by interpreting the modulo(f_s) mapping in the context of a unit circle as shown in Figure 3.5a. The point $1 + j0$ corresponds to the location of all multiples of the sample frequency $f_{1+j0} = kf_s$. The point $-1 + j0$ corresponds to the location of all multiples of the Nyquist frequency $f_{1-j0} = kf_s + f_s/2$. Reconstructed signal frequencies reside on the unit circle with the positive baseband frequencies occupying the top arc (0 to $+180°$), and the negative baseband frequencies occupying the bottom arc (0 to $-180°$). A point on the unit circle at $630°$, for example, is mapped to $270°$ on the unit circle, a point on the negative frequency arc corresponding to $-90°$. The point $-90°$ has a physical value of $f_1 = -f_s/4$, a negative baseband frequency. The EKG data are also analyzed using the circle criteria and Figure 3.5b. For $f_s = 135$ Hz, the aliasing frequency locations are plotted to be ±120 Hz.

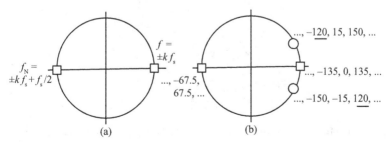

Figure 3.5   (a) Circle criteria used in analyzing $f_1 = f_0 \bmod(f_s)$ where f_1 is a baseband frequency, and (b) location of $f_0 = \pm120$ Hz for a sample rate of 135 Hz and $f_1 = 15$ Hz.

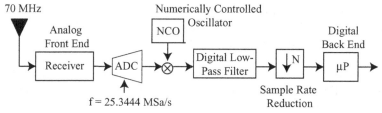

Figure 3.6 Hypothetical RF receiver.

IF SAMPLING

An RF receiver is proposed in Figure 3.6 consisting of an analog RF to intermediate frequency (IF) front end, analog-to-digital converter (ADC), and digital signal processing agents. The signal of interest is centered about a carrier frequency of 70 MHz. According to the sampling theorem, the sample rate must exceed 140 MSa/s. For reasons that may include cost and power dissipation considerations, it is concluded that 100+ MSa/s ADC is unacceptable. A much lower speed (undersampling) and available ADC operating at an f_s = 25.3444 MSa/s rate is preferred. The remainder of the solution is based on a numerically controlled oscillator (NCO) that produces a sine wave at a user-specified frequency, and a digital mixer to heterodyne the sampled signal to baseband where a low-pass filter extracts a selected sub-band from baseband (called a superheterodyne receiver—see insert). Finally, a "back-end" μp decodes received signals.

Major Armstrong (1890–1954)

Factoid: Major Armstrong was an electrical engineer and inventor. He, in fact, invented frequency modulation (FM) radio and the regenerative feedback circuit as an undergraduate (patented in 1914), and the superheterodyne receiver (patented

1918). During World War II, he played a major role in supplying allied troops with communication equipment. Many of Armstrong's inventions were ultimately claimed by others in the form of patent lawsuits. Armstrong found himself embroiled in many patent wars, suffering from unethically and illegal tactics practiced by his adversaries. Alone and depressed over the outcome of the FM patent dispute, Armstrong jumped to his death from the 13th floor window of his New York City apartment on January 31, 1954. His suicide note to his wife said: "May God help you and have mercy on my soul."

Collectively, the system under study is based on what is called an IF undersampling technology. The essential elements of the undersampled receiver are displayed in Figure 3.6. The spectrum of the signal of interest, shown in Figure 3.7 (top), is assumed to be centered about a frequency 70 MHz and is sampled at an ADC rate of 25.344 MSa/s, well below the Shannon sampling rate. The ADC sampling action will therefore produce aliased images of the RF/IF signal that are centered around multiples of the sampling frequencies nf_s, $n = 1, 2$, as shown in Figure 3.7 (middle). It should be noted that the received analog signal also has negative frequency domain components centered at $f = -70$ MHz, which will also produce aliased images centered about nf_s, $n = -1, -2, \ldots$. The analog signal center at 70 MHz will be aliased to a baseband frequency of -6.032 MHz $(70 \times 10^6 - 76.032 \times 10^6)$, while the analog signal center at -70 MHz will be aliased to a baseband frequency of 6.032 MHz $(-70 \times 10^6 + 76.032 \times 10^6)$. The aliased baseband signal will need to be translated down to 0 Hz (DC) to complete the superheterodyne process. This is accomplished by setting the NCO to 6.032 MHz and mixing the ADC and NCO outputs. This will center the original signal located at 70 MHz to a DC location as shown in Figure 3.7 (bottom). From this point, the desired signal can be isolated using a low-pass filter and the information extracted using digital processors.

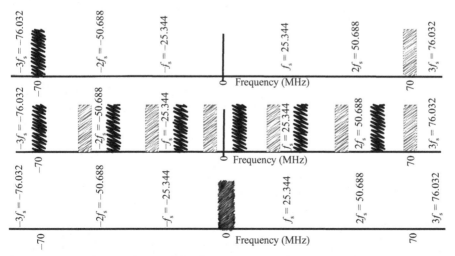

Figure 3.7   Undersampled system showing received RF/IF spectrum (top), sampled spectrum with resulting aliasing (middle), and low-pass filtered spectrum (bottom).

The undersampling story does not end here. While the undersampled strategy resulted in a reduced ADC rate (a modest 25.344 MSa/s vs. 140+ MSa/s), the ADC's sample-and-hold circuit must be capable of producing a nearly ideal (instantaneous) sample of an analog signal of 70+ MHz regardless of the actual ADC sample rate. These devices are called IF sampling ADCs or undersampling ADCs. An example is a commercial 14-bit, 125 MSa/s ADC that has an analog bandwidth of 750 MHz (corresponds to a 1.5 GSa/s ADC). For the design being considered, the 25.344 MSa/s ADC would need to have an analog section with a sampling bandwidth of 140 MHz or better. In addition, the designer would also need to insure that the only RF signal components centered about 70 MHz are presented to the ADC. Any signal outside this range will be aliased back to baseband, producing potentially undesirable results.

Since the presence of aliasing can degenerate the performance of a digital filter, how to manage this phenomenon must be understood. This is particularly true in mixed-signal environments where the signals generated by noisy digital logic can be coupled into the analog section and then sent on an ADC. It has been shown that aliasing can also be creatively used to achieve unexpected solutions such as under-sampled IF systems.

CHAPTER 4

DATA CONVERSION AND QUANTIZATION

DOMAIN CONVERSION

The ubiquitous analog-to-digital converter (ADC) and digital-to-analog converter (DAC) are core digital signal processing (DSP) technologies. In fact, without an ADC or DAC, there would be no DSP today. They are found in applications that involve digital processing of signals that originate and/or terminate in the continuous-time domain. The basic ADC device shown in Figure 4.1 consists of two parts. The first stage performs a sample-and-hold operation that physically converts an analog signal $x(t)$ into a discrete-time signal $x[k]$. The second stage consists of a quantizer that maps the discrete-time sample value $x[k]$ into an equivalent digital word $x_D[k]$. An n-bit ADC quantizes $x[k]$ into one of 2^n possible digital values. In addition, ADCs can be classified to be members of four basic groups. They are

- flash—direct conversion,
- pipelined—segmented direct conversion,
- successive approximation register (SAR)—indirect conversion, and
- sigma-delta ($\Sigma\Delta$)—adaptive conversion.

The mechanism by which a quantized (digital) or discrete-time signal is returned to a continuous-time signal $x(t)$ is called interpolation. The process generally involves the use of a DAC and interpolating filter as shown in Figure 4.2. DACs are commonly configured using a zero-order hold (ZOH) circuit that converts a discrete-time signal into a piecewise constant analog signal $x_Z(t)$. The piecewise constant signal is then "smoothed" by an interpolating filter that produces a continuous-time outcome. DACs, like ADCs, are parameterized in terms of precision (in bits), speed (sample per second or Sa/s), linearity, noise, and other qualifiers. Some DACs come supplied with embedded analog circuits that are used to perform additional signal processing and waveform shaping using modulation and multiplicative operations. Both ADC and DAC technologies are available as a suite of commercial products spanning a wide range of cost, precision, and bandwidth choices.

Digital Filters: Principles and Applications with MATLAB, First Edition. Fred J. Taylor.
© 2012 by the Institute of Electrical and Electronics Engineers, Inc.
Published 2012 by John Wiley & Sons, Inc.

29

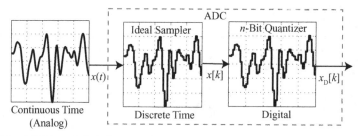

Figure 4.1   ADC-based digital conversion system.

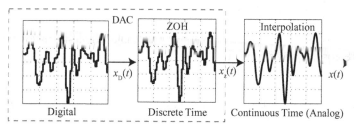

Figure 4.2   DAC-based digital conversion system.

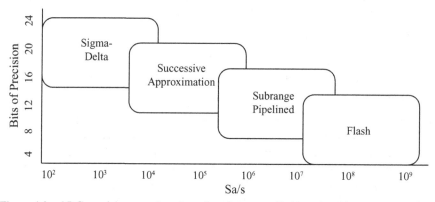

Figure 4.3   ADC precision as a function of performance (Sa/s) and architecture.

ADC TAXONOMY

ADC architectures are varied, resulting in a wide choice of performance, power, cost, and complexities. Choosing the right ADC is fundamentally important to the success of the DSP and filter solution. The general speed versus precision relationships among the basic ADC forms are summarized in Figure 4.3 and comparisons are made in Table 4.1 with the converters graded on a 1–4 scale (1 being the best, 4 worst), with ☑ denoting a capability in the topical area, and ☒ denoting none.

TABLE 4.1. ADC Comparison

Attribute	Flash	Pipelined	SAR	$\Sigma\Delta$
Throughput	1	2	3	4
Resolution (effective number of bits or ENOBs)	4	3	2	1
Latency	1	3	2	4
Suitability for converting multiple signal per ADC	1	2	1	☒
Undersampling	☑	☑	☑	☒
Dithering (increase resolution)	☑	☑	☑	☒

ADC Metrics

Digital data acquisition and conversion systems are found in virtually every modern communication system, DSP solution, electronic instrument and microcontroller applications. As a technology, data conversion systems are seen to evolve at a slow rate compared with mainstream semiconductors. Two data acquisition parameters considered key to many applications are speed and precision. For mobile and untethered applications, a third parameter, power dissipation can be equally important. The parameters that usually define the quality of the ADC conversion process are summarized in Table 4.2.

ADC ENHANCEMENT TECHNIQUES

A typical ADC is approximately linear over its dynamic range. In general, the input signal $x(t)$ may need to be scaled in order to conform to ADC's available dynamic range. This is the role of an automatic gain control (AGC) device. If the input data range exceeds the dynamic range of the ADC, then the ADC will saturate. This action can result in large conversion errors. There are instances, however, when the operating characteristics of an ADC are intentionally made nonlinear in order to reduce the possibility of run-time saturation or dynamic range overflow. To illustrate, the human auditory system is capable of working over 100-dB dynamic range or greater. Man achieves this remarkable capability by logarithmically processing signals. A similar capability can be embedded into an ADC. For example, a signal fully covering a 16-bit range would obviously overwhelm an 8-bit ADC unless the data are first logarithmically compressed as suggested in Figure 4.4. This modification is called companding. It can be seen that large input signals are highly compressed by the compander, and the small-scale signals are amplified. The two common companding strategies in current use are called mu (μ)-law and A-law. Formally, they are

$$\text{mu }(\mu)\text{-law: } y(x) = sign(x)\frac{\log(1+\mu|x|)}{\log(1+\mu)}; \mu = 255 \text{ (typical)} \qquad (4.1)$$

TABLE 4.2.   ADC Parameters

Test Parameter	Unit	Typical Description
Resolution	Bits	The resolution of an n-bit ADC is 1 part in 2^n.
Nonlinearity, differential (DNL)	Bits	Number of bits guaranteed to have no missing codes. Example: 9 bits minimum.
Nonlinearity, integral (INL)	LSB	An ideal ADC operated on a linear operating line ranging from "zero" to "full scale." The maximum deviation from this line is the ADC's integral nonlinearity. Example: ±2 LSB's max.
Full-scale range	V	Difference between the maximum and minimum analog input values specified for the ADC. Example: 0 V to +10 V, single ended; −5 V to +5 V bipolar.
Conversion time	μs	Time required to complete a conversion. Example: 10 μs.
Power supply (±V)	V	ADCs supply voltage. Example: ±5 V.
Signal-to-noise ratio (SNR)	dB	Ratio of measured signal power at the output to internal noise power. $SNR_{max} = 6.02\ N + 1.76$ dB. Example: 45 dB typically at 1 MHz.
Total harmonic distortion (THD)	% or dB	The ratio of the root mean square (rms) sum of the first six harmonic components to the rms value of a full-scale input signal. Example: −88 dB.
Effective number of bits	Bits	Statistical number of bits residing above the noise floor. $ENOB = (SNR_{max} - 1.76)/6.02$.
Spurious-free dynamic range (SFDR)	dB	Amplitude difference between rms value of the fundamental and the largest nonfundamental harmonic spur. $SFDR_{max} = 9\ N - 6$ dBc: Example: −62 dB.
Aperture error	ns	Errors introduced by clock jitter. Example: 1 ns.

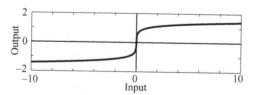

Figure 4.4   Mu-law compander.

and

$$\text{A-law: } y(x) = sign(x) \begin{cases} \dfrac{A|x|}{1+\log(A)}; 0 \le |x| \le (1/A) \\[2ex] \dfrac{1+\log(A|x|)}{1+\log(A)}; (1/A) \le |x| \le 1 \end{cases}. \tag{4.2}$$

DSP DATA REPRESENTATION

In a typical DSP application, data enters a system from an ADC and exits from a DAC. In other applications, the data remains digital from start to finish. In all cases, digital data are manipulated using DSP objects such as a digital filter. DSP solutions are generally multiply-accumulate (MAC) intensive and are often referred to as being SAXPY (S = AX + Y) intense. To achieve high performance and precision, within realistic package, power, and cost constraints, engineers must carefully access their design choices, which collectively define the outcome. One of these decisions relates to the choice of number and/or arithmetic system. In practice, the arithmetic choices are

1. fixed point,
2. floating point, and
3. block floating point (BFP).

Signed fixed-point systems appear in integer or fractional form and can be directly mapped to and from ADCs and DACs. While an unsigned or signed integer representation is familiar to those developing code, the DSP engineer often prefers to work with real numbers (e.g., coefficients and data). To illustrate, consider a 16-bit fixed-point system that consists of a sign bit, 3 integer bits, and 6 bits to fractional bits. Such a system would be said to have $[N:F]$* data format, where $N = 16$ and $F = 6$. A real number $\alpha = 4.2344$ would be coded as the [16,6] word $[0:100 \lozenge 001111]_2 = 4.2344_{10}$ where $\lozenge$ denotes the binary point location. In general, the signed fractional data word X has the form

$$X = \pm \sum_{i=-F}^{I} 2^{-i} X_i; \ X_i \in [0,1]. \tag{4.3}$$

The format associated with a baseline 16-bit Texas Instrument's (TI) DSP processor, for example, is $[16:F]$. TI, however, interprets the data format using what is called "Q" notation and it is $Q(F)$, where $N = 16$ is implicitly assumed. Fixed-point representations, in any of these formats produce only an approximation to a real number α. The difference between α and its fixed-point representation is called the quantization error.

For off-line filtering (e.g., MATLAB), and some real-time applications where precision is a major issue, floating point is preferred. The floating-point representation of a signed number is $x = \pm m_x r^{e_x}$, where m_x is called the floating-point mantissa (generally normalized), r is the radix, and e_x is the exponent. Floating-point formats have been standardized (e.g., IEEE). Their use in digital filtering is justified on the basis of high precision, but they introduce a host of problems as well. Notable is the fact that floating point is slow and resource intense (hardware), and introduces data-dependent latencies that create real-time operational problems.

---

* $[N:F]$ format defines a signed N-bit fixed-point word having a sign bit, F fractional bits of precision, I integer bits, where $I = N - F - 1$.

A variation on the floating-point theme is BFP. Some DSP application-specific standard parts (ASSPs), such as dedicated fast Fourier transform (FFT), use BFP. A BFP system is actually a scaled fixed-point system. Specifically, given an array of data $\{x_i\}$, with maximal element $|x_{max}| \le K = 2^k$, the BFP version of an N-sample array of sample values $\{x_i\}$ is $\{y_i\} = \{x_i/K\} \times 2^k$ where $|x_i/K| \le 1$ and the scale factor 2^k can be reapplied to the data at the end of a computational cycle or, in some cases, ignored altogether. Compared with a fixed-format fixed-point system, a BFP-coded data set can more efficiently utilize the available data dynamic range.

Of these choices, the overwhelming choice for digital filtering is fractional fixed point for hardware-based designs, and specifically 2's complement. The fixed-point advantages in hardware solutions are manifold, including higher speed, reduced complexity, power dissipation, and cost. The fixed-point solution construction often begins with an ADC, which defines, in total or in part, the solution's data format and quantization error. Statistically quantifying this error is necessary in order to rigorously analyze a digital filter.

QUANTIZATION ERROR

Digital signals can be generated by digital devices, such as the ADC shown in Figure 4.5. An ideal impulse sampler instantaneously captures a sample value $x_S[k]$ of an analog signal $x(t)$ at $t = kT_s$. The sampled value is then passed to a quantizer that converts $x_S[k]$ into a digital word $x_D[k]$, where $x_D[k]$ is a digital approximation of $x_S[k]$. The difference between the discrete-time and quantized sample is called the quantization error and is formally defined to be

$$e[k] = x_D[k] - x_S[k]. \tag{4.4}$$

To illustrate, an analog signal $x(t)$ having a ±1-volt swing, shown in Figure 4.6, is sent to an ideal 3-bit signed ADC. The weight of the least significant bit (LSB) is LSB = 1/4 V/bit.

In general, it would be useful to express the ADC errors and properties in terms of computable statistics. If the input signal $x(t)$ is double ended, ranging over $-A \le x[k] < A$, then the quantization step size for an n-bit ADC converter is given by

$$\Delta = 2A/2^n \text{ (double ended)}, \tag{4.5}$$

where Δ is the quantization step size and is normally measured in volts or amps per bit. If the input signal is single ended, ranging over $0 \le x(t) < A$, then the quantization step size for the n-bit converter is given by

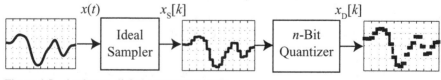

Figure 4.5   Analog-to-digital converter (ADC).

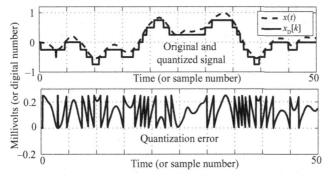

Figure 4.6 Digitized analog signal using a 3-bit ADC (top) and the resulting quantization error (bottom).

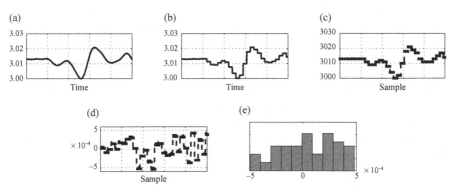

Figure 4.7 Production of quantization errors. (a) Original analog signal; (b) sampled analog signal; (c) digitized signal; (d) quantization error; (e) PDF for quantization noise.

$$\Delta = A/2^n \text{ (single ended)}, \quad (4.6)$$

which is ½ the double-ended quantization step size. For linear ADC applications, there is a straightforward statistical error analysis procedure that can quantify the error process. The analysis process is based on the production of the quantization error, which can be expressed in terms of the quantization step size, Δ (also commonly denoted as Q). To illustrate, assume that a signed (doubled ended) 16-bit ADC is designed to operate over the analog range ± 32.768 V. The ADC quantization step size is given by $\Delta = 2 \times 32.768/2^{16} = 10^{-3}$ (V/bit). A sampled value of 3.015 V, shown in Figure 4.7, would be mapped to the quantized integer value of 3015 without quantization error. The signal being quantized in Figure 4.7 is known to be locally distributed randomly about 3.015 V in time. The resulting quantized values are therefore distributed about 3015. Using rounding, a particular sample value, say $x_S[k] = 3.0152$ V, would be mapped to an integer value 3015, which corresponds to a real value $x_D[k] = 3.015 = 3015\Delta$. The quantization error, in this case, would be $e[k] = 3.015 - 3.0152 = -0.0002$. The quantization error, shown on a sample-by-sample basis is seen to reside within the range $[-\Delta/2, \Delta/2)$ (LSB weight $= \Delta$), which

TABLE 4.3.   Quantization Error Statistics

Policy	Figure	Maximum |Error|	Error Range	Error Mean	Error Variance
Truncation	16(a)	ΔLSB	$e[k] \in [0, \Delta)$	$\Delta/2$	$\Delta^2/12$
Rounding	16(b)	$\Delta/2$	$e[k] \in [-\Delta/2, \Delta/2)$	0	$\Delta^2/12$

corresponds to a maximum ±0.5-mV error. The error statistics can be defined in terms of the error probability density function or PDF. The PDF, when computed and plotted as a histogram, is seen to have a "blocky" distribution over the range ±½ LSB as shown in Figure 4.7. If the experiment had been run over a longer sample record, and if the histogram resolution was increased, the PDF would converge to a uniform distribution.

A quantizer can be set to perform rounding or truncation operations. If an ADC's word width is n-bits, and if n is sufficiently large (typically $n \geq 4$ bits), the quantization error $e[k]$ is generally assumed to be a uniformly distributed random process. Specifically, the statistical error model for rounding is a uniformly distributed PDF centered about zero (i.e., zero mean). For truncation, the error model is again uniform, but the mean error is now $\Delta/2$. The first and second moment statistics (i.e., mean and variance), as well as the maximum error for each case are shown in Table 4.3.

The data shown in Table 4.3 state that when the quantization error is uniformly distributed, the error variance is

$$\sigma^2 = \frac{1}{\Delta} \int_{-\Delta/2}^{\Delta/2} x^2 dx = \Delta^2/12. \qquad (4.7)$$

The variance can also be interpreted as a standard deviation given by $\sigma = \Delta/\sqrt{12}$ having units of Δ (e.g., volts per bit). The error, in bits, can be expressed in bits as

$$\text{error (in bits)} = \log_2(\sigma) = \log_2(\Delta) - 1.79 \qquad (4.8)$$

or in decibels as error (dB) = $20\log_{10}(\sigma) = (20\log_{10}(\Delta) - 10.8)$ dB.

Example: Error Analysis

The input to a signed 16-bit double-ended ADC is assumed to be bounded by $|x(t)| \leq 1$. The quantization step size is therefore $\Delta = 2^{-15}$ and an attendant error variance is $\sigma^2 = \Delta^2/12 = 2^{-30}/12$. The test signal, an array of random variables, is quantized into fractional numbers having m-bits of fractional precision, where m ranges from 1 to 15 bits. The outcome is graphically interpreted in Figure 4.8. It can be seen that there is an agreement between the theoretical prediction and the experimental outcome.

Drilling deeper into the details of the quantization model, consider a double-ended analog input signal $x(t)$ bounded by ±10 V is to be digitized using a ±10 V ($A = 10$) 10-bit signed ADC. While the errors can be readily predicted and computed,

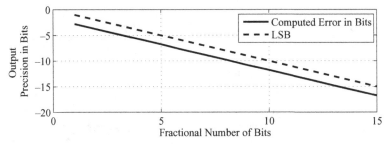

Figure 4.8   Computer simulation outcome displaying the quantization error (in bits) of a random input for a fractional precision m ranging from 1 to 15 bits. The experimental data clearly illustrate about 1.79-bit bias in the error statistics.

a common problem is interpreting these errors. From the given data, the quantization step size is computed to be $\Delta = 20/1024 \sim 20$ mV/bit, a number that can be used to calibrate the statistical error study. The ADC output corresponds to a 10-bit data word having 1 sign bit and 9 data bits, where $2^9*\Delta = 512 \times 0.02 \sim 10$ (single-ended range limit). The quantization error $e[k] = x_S[k] - x_D[k]$ is assumed to be uniformly distributed over $[-\Delta/2, \Delta/2)$. The resulting rounded error statistics, defined in Table 4.3, are $E(e) = 0$ and $\sigma^2 = \Delta^2/12$, or $\log_2(\sigma) = 7.43$ bits. A logical question to ask is what does -7.43 bits of precision actually mean? The quantization error, in bits, was computed using the formula $\log_2(\sigma) = \log_2(\Delta) - 1.79$ bits and has a value of -7.43 bits and is statistically interpreted as 7.43 fractional bits of ADC precision. Removing the statistical bias of 1.7 bits from the computed quantization error -7.43 results in an absolute ADC precision of $F = 5.63$ fractional bits. Note that $25.63 \times \Delta = 25.63 \times 0.02 = 0.990442 \sim 1.0$, which covers ADC's fractional dynamic range $[0,1]$. Subtracting the number of fractional bits $F = 5.63$ bits from the 10-bit data format, one concludes that the number of retained integer bits (i.e., I) per sample satisfies $10 = I + F + 1$ or $I = 3.4$. Observe that $2^{3.4} = 9.849 \sim 10$ is the limit of the single-ended ADC dynamic range. The data format would code data using a $[\pm:$ $I \lozenge F]$ format, where $\lozenge$ denotes the binary point, I is the number of integer bits, and F is the fractional word width in bits. The system under study would technically have the data format $[\pm: 3.4 \lozenge 5.6]$. In "DSP speak," the data format would be denoted $[N:F] = [10:5.6]$, and in "TI speak," the format would be $Q(5.6)$ where $N = 16$ is implicitly assumed.

In the time domain, the quantization error is assumed to be a uniformly distributed random process. It is normally assumed that in the frequency domain, the noise spectrum is "flat" over the baseband range $f \in [0, f_s/2)$. Specifically, a critically sampled ADC produces a quantization noise process that is uniformly distributed over the frequency range $f \in [0, f_s/2)$. If the ADC is oversampled by a factor k, the same noise power is distributed over $f \in [0, kf_s/2)$, which reduces the density of noise by a factor k as suggested in Figure 4.9. A low-pass filter can then be used to eliminate the noise contributions beyond the Nyquist frequency $f_s/2$, leaving only $1/k$th the original noise power located at baseband.

A basic $\Delta\Sigma$ ADC, by its very nature, is a highly oversampled system. In addition, the $\Delta\Sigma$ ADC uses an integrator, having a frequency response $H_I(f) = 1/f$ to

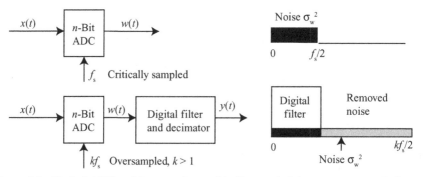

Figure 4.9   Typical ADC architectures (top: critically sampled; bottom: oversampled).

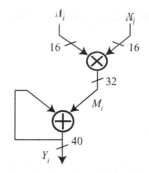

Figure 4.10   Basic MAC unit showing data bus widths of 16, 32, and 40 bits.

process internal error signals. As a result, the noise found at the higher frequencies is rejected by the integrator, resulting in a higher than expected effective number of bits (ENOBs) of precision.

MAC UNITS

Quantization errors can occur whenever and wherever real data are replaced by a fixed-point equivalent. Besides ADCs, there are other quantization error sources found in a fixed-point system. It has been established that DSP solutions are SAXPY-intensive, which implies that there exists arithmetic units, including multipliers, that can introduce additional quantization errors. Managing and measuring such errors is both necessary and challenging. To illustrate, suppose two signed real data arrays A and X consisting of 256 samples each are to be combined to form a sum-of-products (SOPs) outcome that can be expressed as

$$Y = \sum_{i=0}^{255} A[i] X[i] \tag{4.9}$$

and graphically interpreted in Figure 4.10. Figure 4.10 indicates that there are different word lengths assigned to data at different locations within the MAC unit.

Assume that the $N = 16$ bit versions of $|A_i| \le 1$ and $|X_i| \le 1$ are quantized so that $A \; \varepsilon \; [16{:}15]$ and $X \; \varepsilon \; [16{:}15]$. The full precision product M_i of each A_i and X_i pair is then modeled to be a 32-bit word requiring a format of $M \; \varepsilon \; [32{:}31]$. To insure that the output register of the accumulator does not overflow during run time, it must survive a worst case attack. The worst case SOP $|Y|$ is 256, which is the result of accumulating 256 worst case individual products. Since this case can occur during run time, the developers of DSP hardware technology have made provision in designing MAC units so that they can operate under worst case conditions without interruption. Specifically, it is recognized that the output register containing Y must be able to accept data outcomes out to 256, which is 8 integer bits larger than the individual multiplier data word. This is accomplished by designing the accumulator data paths 8 bits wider than the full precision multiplier data paths. This is called "extended precision" accumulator having an output $Y \; \varepsilon \; [32 + 8{:}31] = [40{:}31]$. If it were not for such dynamic range extensions, register overflows would likely occur during run time, resulting in either large and disturbing errors or excessive overhead in constantly checking for register overflow and correcting it when it occurs. Once the SAXPY SOP has been completed, the extended precision outcome $Y \; \varepsilon \; [40{:}31]$ can be rounded to a 16-bit word having a $[16{:}7]$ format. The rounded outcome therefore would statistically have $-7 - 1.8 = -8.8$ bits of fractional precision.

MATLAB SUPPORT

MATLAB contains several tools that can be used to support quantization studies. They are

- *ceil* function: round toward ∞,
- *floor* function: round toward $-\infty$,
- *fix* function: round toward 0,
- *round* function: round toward nearest integer, and
- *hist* function: histogram.

Other measures include

- *median, std, min, max, var.*

CHAPTER 5

THE Z-TRANSFORM

INTRODUCTION

Like so many technical and scientific fields, digital signal processing (DSP) has developed its own language and idioms. The language of DSP is rooted in the mathematics that describes the objects under study. For digital filters, this is the z-transform, a bridge between continuous-time (analog) systems and their discrete-time counterparts. Discrete-time systems, in turn, provide the framework upon which digital filters are built. The origins of the z-transform can be traced to the Laplace transform. In studying continuous-time signals and systems, it was soon realized that the analysis complexity could be significantly reduced by replacing calculus operations with Laplace-defined algebraic statements. The Laplace transform, in a sense, transformed 20th-century academic engineering, placing the study of linear

Pierre-Simon Marquis de Laplace (1749–1827), French mathematician and astronomer.

Digital Filters: Principles and Applications with MATLAB, First Edition. Fred J. Taylor.
© 2012 by the Institute of Electrical and Electronics Engineers, Inc.
Published 2012 by John Wiley & Sons, Inc.

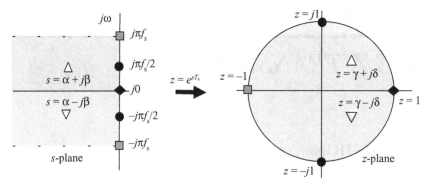

Figure 5.1   Mapping of the s-plane into the z-plane under $z = e^{sT_s}$.

systems on a new mathematical foundation. When discrete-time signals first made their appearance in the mid-20th century as sampled-data control systems, they were initially studied in the context of Laplace transforms using a well-known delay property of Laplace transforms. The delay theorem states that if $x(t) \leftrightarrow X(s)$, then $x(t - kT_s) \leftrightarrow e^{-skT_s} X(s)$. As a result, a time series $x[k] = \{x[0], x[1], x[2], \ldots\}$ can be alternatively represented using the Laplace transform and the delay theorem as $X(s) = x[0] + x[1]e^{-sT_s} + x[2]e^{-2sT_s} + \cdots$. The fundamental problem with the Laplace representation methodology is that each delay term requires the insertion of a delay operator of the form e^{-skT_s}. For periodically sampled signals, an infinite number of insertions would be required. Obviously, this can become very tedious and rapidly gave rise to the adoption of a shorthand representation of the form

$$z \triangleq e^{sT_s} \quad \text{or} \quad z^{-1} \triangleq e^{-sT_s}, \tag{5.1}$$

which is known by its popular name the z-operator and z-transform. Since its introduction, the venerable Laplace s-operator, and more recently the z-operator, has become a ubiquitous tool in the study of continuous- and discrete-time linear signals and systems.

Equation 5.1 defines a relationship between points in the s-plane and those in the z-domain as graphically shown in Figure 5.1. The z-transform is seen to be a mapping of the complex Laplace operator $s = \sigma + j\omega$, into the z-domain as $z = e^{sT_s} = e^{\sigma + j\omega} = re^{j\omega}$. For $\omega = k(2\pi) + \omega_0$, $z = re^{j(k2\pi + \omega_0)} = re^{j\omega_0}$. It can therefore be concluded that if the z-transform mapping is to be unique, the imaginary part of s must be restricted to a range $\omega \in [-\pi, \pi)$. This corresponds to limiting the frequency range to be only baseband frequencies, physically residing between the $\pm f_{\text{Nyquist}}$ in a manner consistent with Shannon's sampling theorem.

Z-TRANSFORM

The two-sided (bilateral) z-transform of an arbitrary time series $x[k]$ is formally defined to be

$$X(z) = \sum_{k=-\infty}^{\infty} x[k]z^{-k}. \tag{5.2}$$

For a causal time series, the one-sided (right-sided, unilateral) z-transform applies and is given by the right-sided, or one-sided summation:

$$X(z) = \sum_{k=0}^{\infty} x[k]z^{-k}. \tag{5.3}$$

A time series may also be of finite duration. In such cases, the z-transform is given by the finite sum

$$X(z) = \sum_{k=m}^{n} x[k]z^{-k}. \tag{5.4}$$

The values of z for which the sums found in Equations 5.2 through 5.4 produce a bounded outcome are called the z-transform's region of convergence or ROC. The importance of the ROC to the study of discrete-time signals is subject to debate. Some place a great emphasis on establishing the ROC of a transform, others totally ignore this study altogether. Whether the ROC is emphasized or deemphasized, the production of the z-transform of an arbitrary signal $x[k]$ can be a challenging problem. Fortunately, the z-transform for many important time series (i.e., elementary functions) is well known and has been cataloged.

PRIMITIVE SIGNALS

In general, the z-transform of simple signals is rarely derived from first principles. Instead, they are synthesized using elements from standard tables (e.g., Table 5.1) and known properties of z-transforms (e.g., linearity). A collection of these signals, their z-transforms, and ROCs are shown in Table 5.1.

Z-TRANSFORM: LINEAR SYSTEMS

Digital filters can be modeled and analyzed in the z-domain. A common means of analyzing a filter is to present the filter with a Kronecker delta function $\delta_K[k]$ as an input and to analyze the outcome (i.e., impulse response). From such studies, the response of a filter to an arbitrary input can be derived. In general, the input–output behavior of an at-rest (zero initial conditions) linear constant coefficient system, or filter, can be characterized by the difference equation:

$$\sum_{m=0}^{N} a_m y[k-m] = \sum_{m=0}^{M} b_m x[k-m], \tag{5.5}$$

TABLE 5.1. *z*-Transforms of Elementary Functions

Time Domain	*z*-Transform	Region of Convergence
$\delta[k]$	1	Everywhere
$\delta[k-m]$	z^{-m}	Everywhere
$u[k]$	$z/(z-1)$	$|z| > 1$
$ku[k]$	$z/(z-1)^2$	$|z| > 1$
$k^2u[k]$	$z(z+1)/(z-1)^3$	$|z| > 1$
$k^3u[k]$	$z(z^2+4z+1)/(z-1)^4$	$|z| > 1$
$a^ku[k]$	$z/(z-a)$	$|z| > |a|$
$ka^ku[k]$	$az/(z-a)^2$	$|z| > |a|$
$k^2a^ku[k]$	$az(z+a)/(z-a)^3$	$|z| > |a|$
$\sin[bk]u[k]$	$\dfrac{z\sin(b)}{z^2-2z\cos(b)+1}$	$|z| > 1$
$\cos[bk]u[k]$	$\dfrac{z(z-\cos(b))}{z^2-2z\cos(b)+1}$	$|z| > 1$
$\mathrm{Exp}[akT_s]\sin[bkT_s]u[kT_s]$	$\dfrac{ze^{aT_s}\sin(bT_s)}{z^2-2ze^{aT_s}\cos(bT_s)+e^{2aT_s}}$	$|z| > |\exp(aT_s)|$
$\mathrm{Exp}[akT_s]\cos[bkT_s]u[kT_s]$	$\dfrac{z(z-e^{aT_s}\cos(bT_s))}{z^2-2ze^{aT_s}\cos(bT_s)+e^{2aT_s}}$	$|z| > |\exp(aT_s)|$
$a^k\sin(bkT_s)u[kT_s]$	$\dfrac{az\sin(bT_s)}{z^2-2az\cos(bT_s)+a^2}$	$|z| > |a|$
$a^k\cos(bkT_s)u[kT_s]$	$\dfrac{z(z-a\cos(bT_s))}{z^2-2az\cos(bT_s)+a^2}$	$|z| > |a|$
$a^k(u[k]-u[k-N])$	$(1-a^Nz^{-N})/(1-az^{-1})$	Everywhere

where $y[k]$ is the output at the kth sample instance and $x[k]$ is the input also at the kth sample instant. Suppose the z-transform of $y[k]$ is denoted $Y(z)$, and the z-transform of $x[k]$ is denoted $X(z)$. Then, it immediately follows from the delay theorem that $y[k-m] \leftrightarrow z^{-m}Y(z)$ and $x[k-m] \leftrightarrow z^{-m}X(z)$, which results in the z-transformed difference equation given by

$$\left(\sum_{m=0}^{N} a_m z^{-m}\right)Y(z) = \left(\sum_{m=0}^{M} b_m z^{-m}\right)X(z). \tag{5.6}$$

The ratio of $Y(z)$ to $X(z)$ is traditionally called a transfer function and is denoted $H(z)$. For the linear system, described by Equation 5.6, the transfer function is formally given by

$$H(z) = \frac{Y(z)}{X(z)} = \frac{\left(\displaystyle\sum_{m=0}^{M} b_m z^{-m}\right)}{\left(\displaystyle\sum_{m=0}^{N} a_m z^{-m}\right)}. \tag{5.7}$$

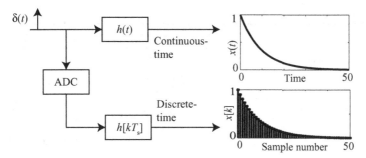

Figure 5.2   Impulse invariant system response.

Equation 5.7 describes how the z-transform of the input signal is "transformed" to the z-transform of the output signal. Much information about a system is stored or embedded in the transfer function.

One of the useful attributes of the z-transform is called the impulse invariant property as shown in Figure 5.2. This can be illustrated by considering a simple first-order continuous-time system $H(s) = 1/(s + 1)$ having an impulse response $h(t) = e^{-t}u(t)$. For $T_s = 0.1$, the pole at $s = -1$ is mapped to $z = e^{-sT_S} = e^{-0.1} = 0.9048 = a$. Therefore, from Table 5.1, $H(z) = z/(z - a)$ and an impulse response $h[k] = a^k u[k]$, which corresponds to $h(t)\big|_{t=kT_s} = e^{-t}u(t)\big|_{t=kT_s} = a^k u[k]$. That is, the continuous- and discrete-time impulse responses are identical at the sample instances $t = kT_s$, which is predicted by the impulse invariance property.

Z-TRANSFORMS PROPERTIES

It is generally assumed that most of the important signals can be represented as a collection of one or more elementary functions found in Table 5.1. However, these primitive signals may need to be altered in some form using one or more of the operators, or modifiers, listed in Table 5.2. The operations listed in Table 5.2 describe how complicated signals can be constructed by manipulating and combining primitive signals.

In addition to the properties listed in Table 5.2, there are several other operators that need mentioning. The first is called the initial value theorem, which states that

$$x[0] = \lim_{z \to \infty} X(z) \tag{5.8}$$

if $x[k]$ is causal. The second property is called the final value theorem and is given by

$$x[\infty] = \lim_{z \to 1}(z-1)X(z) \tag{5.9}$$

provided that the poles of $(z - 1)X(z)$ are interior to the unit circle (i.e., $|z| < 1.0$).

TABLE 5.2. Properties of z-Transforms

Property	Time Series	z-Transform
Homogeneity (scaling)	$\alpha x[k]$	$\alpha X(z)$
Additivity	$x_1[k] + x_2[k]$	$X_1(z) + X_2(z)$
Linearity	$\alpha x_1[k] + \beta x_2[k]$	$\alpha X_1(z) + \beta X_2(z)$
Left shift (single delay)	$x[k + 1]$	$zX(z) - x[0]$
Left shift (multiple delays)	$x[k + N]$	$z^N\left(X(z) - \sum_{i=0}^{N-1} x[i]z^{-i}\right)$
Right shift (multiple delays)	$x[k - N]$	$z^{-N}X(z)$
Complex conjugation	$x^*[k]$	$X^*(z^*)$
Reversal	$x[-k]$	$X(1/z)$
Complex modulation	$e^{j\omega k}x[k]$	$X(e^{-j\omega}z)$
Multiplication by a complex power series	$w^k x[k]$	$X(z/w)$
Ramping	$kx[k]$	$-z\dfrac{dX(z)}{dz}$
Reciprocal decay	$\dfrac{1}{k}x[k]$	$-\oint \dfrac{X(\zeta)}{\zeta}d\zeta$
Summation (accumulation)	$\sum_{n=-\infty}^{k} x[n]$	$\dfrac{zX(z)}{(z-1)}$
Periodic extension of signal of length N	$\sum_{n=0}^{\infty} x(k+nN)$	$\dfrac{z^N X(z)}{z^N - 1}$
Convolution	$x_1[k]x_2[k]$	$X_1(z)X_2(z)$
Correlation	$x_1[k]x_2[-k]$	$X_1(z)X_2(z - 1)$
Parseval's equation	$\sum_{k=-\infty}^{\infty} x_1[k]x_2^*[k]$	$\dfrac{1}{j2\pi}\oint X_1(\sigma)X_2^*\left(\dfrac{1}{\sigma^*}\right)\sigma^{-1}d\sigma$

MATLAB Z-TRANSFORM SUPPORT

MATLAB's Symbolic Math Toolbox provides tools for solving and manipulating symbolic math expressions. The toolbox contains symbolic functions that are expressed in MATLAB's MuPAD language. All functions can be accessed from the MATLAB command line or from the MuPAD notebook interface. A tool having particular interest at this time is ztrans, which returns the z-transform from a given expression of $x[k]$. To illustrate, consider analyzing $x[k] = a^k u[k]$, $|a| < 1$, as found in Table 5.1, using MATLAB:

```
» syms a n % declares the symbolic variables

» x= a^n;

» X=ztrans(x);
```

```
» X

z/a/(z/a-1)
```

The result is $X(z) = (z/a)/((z/a) - 1) = z/(z - a)$. The MATLAB production of the z-transform of a sinusoid of the form $x[k] = \sin(ak)u[k]$, as found in Table 5.1, proceeds as follows:

```
» syms a n % declares the symbolic variables

» x=sin(a*n);

» X=ztrans(x);

» X

z*sin(a)/(z^2-2*z*cos(a)+1)
```

The result is $X(z) = \sin(a)z/(z^2 - 2z\cos(a) + 1)$, $|a| < 1$. There are times, however, when interpreting the MATLAB outcome can be challenging.

SYSTEM STABILITY

A system is classified as being bounded-input bounded-output stable (BIBO) if the output remains bounded for all possible bounded inputs. The z-transform, as defined by Equation 5.1, provides a simple means of classifying the stability of a linear constant coefficient (time-invariant) discrete-time system. Specifically, $z = e^{sT_s}$ maps the poles of a continuous-time system to the z-plane. The stable pole locations of a continuous-time system can then be translated into stable pole locations of a discrete-time system. The mapping of the pole locations of a continuous-time system into the z-plane is graphically shown in Figure 5.3 and summarized in Table 5.3. From classic continuous-time linear system theory, poles can be classified as being asymptotic stable, conditionally stable, and unstable depending on the pole locations. For example, a continuous-time system $H(s) = N(s)/D(s)$ is asymptotic stable if all the poles are located in the left-hand s-plane. Under the z-transform, the left-hand

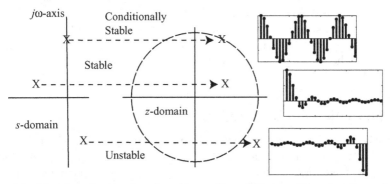

Figure 5.3   Mapping of s-domain poles with specific stability conditions into the z-plane.

TABLE 5.3. Pole Locations of Continuous- and Derivative Discrete-Time Systems

Stability Class	s-Domain	z-Domain	Example	Pole Locations
Asymptotically stable	$\mathrm{Re}(s) < 0$	$\|z\| < 1$ (interior to the unit circles)	$H(z) = z/(z - \alpha),$ $\|\alpha\| < 1$	$z = \alpha,$ $\|\alpha\| < 1$
Conditionally stable	At most one pole at any point on the $j\omega$-axis with all other poles in the left-hand plane	At most one pole is on the unit circle with all other poles being interior to the unit circle	$H(z) = z^2/(z - 1)$ $(z - \alpha),$ $\|\alpha\| < 1$	$z = 1,$ $z = \alpha,$ $\|\alpha\| < 1$
Unstable	More than one pole at a point on the $j\omega$-axis or at least one pole satisfying $\mathrm{Re}(s) > 0$	More than one pole at a point on the unit circle or at least one pole outside the unit circle	$H(z) = z^2/(z - \alpha^{-1})$ $(z - \alpha),$ $\|\alpha\| < 1$	$z = 1/\alpha,$ $z = \alpha,$ $\|\alpha\| < 1$

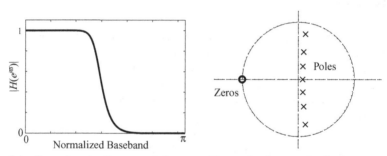

Figure 5.4   Seventh-order Butterworth low-pass filter and pole-zero distribution.

s-plane poles are mapped to points interior to the unit circle in the z-plane. The production of the pole locations associated with a given transfer function $H(z) = N(z)/D(z)$ can be facilitated by using a general purpose math analysis software tool, such as MATLAB. Specifically, the MATLAB `roots` command can be used to factor the nth-order polynomial $D(z)$ to produce a list of poles $z = p_i$. By evaluating the magnitude of the roots ($|p_i|$), the pole's relative position on, interior, or exterior to the unit circle in the z-domain can be determined and the stability class can be indentified.

Example: Pole-Zero Distribution

MATLAB can be used to define the transfer function of a seventh-order Butterworth low-pass filter having a passband cutoff frequency of $f = 0.225 f_s$ as shown in Figure 5.4 along with the filter's pole-zero distribution. The poles are computed and found to consist of three complex-conjugate pole pairs plus one real pole. Their absolute values are all bounded below unity, which insures that the poles are interior to the unit circle guaranteeing stability:

MATLAB code

```
»  [B,A]=butter(7,0.45);  [H,w]=freqz(B,A);  plot(w,abs(H))

»  zplane(B,A);  roots(A)

   0.1282 + 0.7894i;  0.1282 - 0.7894i

   0.0968 + 0.4779i;  0.0968 - 0.4779i

   0.0828 + 0.2268i;  0.0828 - 0.2268i

   0.0787

»  abs(roots(A))

   0.7998;  0.7998

   0.4876;  0.4876

   0.2414;  0.2414

   0.0787
```

INVERSE Z-TRANSFORM

The z-transforms of primitive signals, such as the exponential $x[k] = a^k u[k]$, have generally been reduced to table entries (e.g., Table 5.1). From a knowledge of $X(z)$, it is possible to recover the original discrete-time signal $x[k]$ using a process called

Oliver Heaviside (1850–1925), English electrical engineer.

the inverse z-transform. The most popular inverse z-transform instantiation is called the partial fraction expansion or the Heaviside expansion method.

HEAVISIDE EXPANSION METHOD

The Heaviside method is based on a partial fraction expansion of $X(z)$ and takes the form

$$X(z) = \frac{N(z)}{D(z)} = \alpha_0 + \ldots + \frac{\alpha_{j,n_j} z}{(z - \lambda_j)^{n_j}} + \ldots + \frac{\alpha_{j,k} z}{(z - \lambda_j)^k} + \ldots + \frac{\alpha_{j,1} z}{(z - \lambda_j)} + \ldots \quad (5.10)$$

It can be observed that all the terms in Equation 5.10 are in one-to-one correspondence with the terms found in a standard z-transform table (e.g., Table 5.1). The coefficients $a_{i,k}$ are called Heaviside coefficients and correspond to the eigenvalue λ_i appearing with multiplicity n_i. The Heaviside coefficient production rules are complicated, especially for high-order systems with repeated roots. The production formula are shown below for an eigenvalue λ_j appearing with multiplicity n_j:

$$\alpha_{j,n_j} = \lim_{z \to \lambda_j} \left(\frac{(z - \lambda_j)^{n_j} X(z)}{z} \right), \quad (5.11)$$

$$\alpha_{j,k} = \lim_{z \to \lambda_j} \left(\frac{1}{(n_j - k)!} \frac{d^{(n_j - k)}}{dz^{(n_j - k)}} \left\{ \frac{(z - \lambda_j)^{n_j} X(z)}{z} \right\} \right); \quad k \in [1, n_j - 1]. \quad (5.12)$$

Equations 5.11 and 5.12 can be manually computed, but the process can be exasperating and subject to computational errors. In a contemporary setting, a general purpose computer is normally used to compute the Heaviside coefficients. The MATLAB functions `residue` and `residuez` can be used to compute Heaviside coefficients. The function `residue` is used when working with polynomials $X(z)$ in ascending power and `residuez` is used in the descending case. The name residue is derived from residue calculus, which provides a theoretical foundation for z-transforms.

Example: Heaviside Expansion

Suppose $X(z) = z^2/[(z - 1)(z - 0.5)]$, where $x[k]$ is causal. It is apparent that the poles of $X(z)$ are distinct (nonrepeated) and are located at $z = 1$ and $z = 0.5$. The general form of the partial fraction expansion of $X(z)$ is

$$X(z) = \frac{z^2}{(z-1)(z-0.5)} = \alpha_0 + \alpha_1 \left(\frac{z}{z-1} \right) + \alpha_2 \left(\frac{z}{z-0.5} \right) = 0 + 2 \left(\frac{z}{z-1} \right) - \left(\frac{z}{z-0.5} \right)$$

having an inverse z-transform $x[k] = (2 - 0.5^k)u[k]$ according to Table 5.1. Using MATLAB and the `residue` command, the following results are obtained when expanding $X(z)/z$ (not $X(z)$) in ascending powers of z:

```
» b=[1,0,0];    %  N(z)

» a=[1,-1.5,0.5,0];    &  zD(z)

» residue(b,a)  %  invert  N(z)/zD(z)

ans =       {Heaviside  coefficients}

      2     {α₁}

     -1     {α₂}

      0     {α₀}
```

which agrees with the previously presented outcome. If $X(z)$ is written in descending order, `residuez` can be used. As before,

$$X(z) = \frac{z^2}{(z-1)(z-0.5)} = \frac{1}{(1-z^{-1})(1-0.5z^{-1})}$$

$$= \alpha_1 \frac{1}{(1-z^{-1})} + \alpha_2 \frac{1}{(1-0.5z^{-1})} = \alpha_1 \frac{z}{(z-1)} + \alpha_2 \frac{z}{(z-0.5)}.$$

Then, using MATLAB's *residuez* the Heaviside coefficients α_1 and α_2 are obtained as follows:

```
» b=[1];    %  N(z^-1)

» a=[1,-1.5,0.5];    %  D(z^-1)

» residuez(b,a)

ans =

      2

     -1
```

which results in the previously stated solution.

MATLAB INVERSE Z-TRANFORM SUPPORT

MATLAB's Symbolic Math Toolbox can facilitate the inversion of a given $X(z)$. The tool `iztrans` returns an inverse z-transform of a given $X(z)$. To illustrate, consider inverting $X(z) = z/(z - 0.5)$:

```
» syms z k

» X=z/(z-0.5)

» x=iztrans(X,z,k)

X = (1/2)^k
```

or $x[k] = (0.5)^k u[k]$. Consider now $X(z) = z^2/(z^2 - 1.5z + 0.5)$:

```
» syms z k
» X=(z^2)/(z^2 -1.5*z +0.5)
» x=iztrans(X,z,k)
x =2-(1/2)^k
```

or $x[k] = (2 - (0.5)^k)u[k]$. Finally, consider $X(z) = z^2/(z - 1)^2$:

```
» syms z k
» X=(z^2)/(z-1)^2
» x=iztrans(X,z,k)
x =1+k
```

or $x[k] = (1 + k)u[k]$. It should be noted, however, that for higher-order instances, the symbolic inverse z-transform delivered can be difficult to interpret.

CHAPTER **6**

FINITE IMPULSE RESPONSE FILTERS

INTRODUCTION

Electronic filters are generally assumed to be defined as continuous-time, discrete-time, or digital devices. Signals in these domains appear in one, two, or multiple dimensions. Some signals are completely parameterized by their time and/or frequency domain attributes; others are defined in some statistical sense. The function of an electronic filter is to alter or manipulate a signal's shape, energy, distributions, and other attributes in some predetermined manner. Filters, for example, can be created to alter audio and video records, transforming signals into forms that conform to the designer's personal tastes and preferences. In communication applications, filters are used to detect and select signals of interest that reside within a prespecified frequency band, to suppress noise, and to correct for imperfections in a communication channel. Filters can be used to monitor human health as well as the health of other machines. The list of applications is virtually endless.

Analog, or continuous-time filters, have been part of man's world since the dawn of time. They are part of the human auditory system. Man has also fashioned optical filters in the form of lens to sharpen images, and developed hydraulic filters, called shock absorbers, to smooth an uneven roadway. Beginning in the first half of the 20th century, analog electronics gave rise to a new class of filters based on resistors, capacitors, inductors, and amplifiers. The outcome was radio, television, and other electronic wonders. The mid-20th century saw a short-lived era of discrete-time or sample data filters. This technology rapidly gave way to the products of the digital revolution, an era we now inherit. Digital technology has ushered into existence a plethora of digital hardware, firmware, and software-defined filter solutions and products. In many cases, digital filters are used as an analog replacement technology. In other cases, digital technology has enabled new filters and filter applications that previously never existed. Digital filters have now matured to point where they exhibit a long list of attributes including

- high-precision and accuracy,
- programmability and adaptability,

Digital Filters: Principles and Applications with MATLAB, First Edition. Fred J. Taylor.
© 2012 by the Institute of Electrical and Electronics Engineers, Inc.
Published 2012 by John Wiley & Sons, Inc.

- precise phase and latency control,
- robust performance over a wide range of frequencies,
- compact size and interoperability with other digital subsystems,
- low-cost and low-power dissipation, and
- high reliability and repeatability.

Baseline digital filters have evolved along two major paths called finite impulse response (FIR) filters and infinite impulse response (IIR) filters. The simplest and, in many cases, the most important of these filter classes is the FIR.

FIR FILTERS

As the name implies, an FIR filter has finite duration impulse response. The impulse response of an Nth-order FIR is defined by the N-sample time series $h_N[k] = \{h[0], h[1], \ldots, h[N-1]\}$. A common implementation of the filter is shown in Figure 6.1. This form, called a direct FIR architecture, consists of a collection of shift registers configured as a first-in first-out (FIFO) array, N multipliers, and an accumulator. Observe that if the filter's input is an impulse $x[k] = \delta[k]$, then the FIR's output time series $y[k]$ is $\{h[0], h[1], \ldots, h[N-1]\}$, the filter's impulse response. Mathematically, the output response of an Nth-order linear FIR, having an impulse response $h[k]$ to an arbitrary input time series $x[k]$, is defined by the linear convolution sum

$$y[k] = \sum_{m=0}^{N-1} h[m] x[k-m] = \sum_{m=0}^{N-1} h[k-m] x[m]. \tag{6.1}$$

In z-transform domain, the filter's transfer function can be expressed as

$$H(z) = \sum_{k=0}^{N-1} h_k z^{-k}. \tag{6.2}$$

The filter's frequency response can be obtained by evaluating $H(z = e^{j\varpi})$ over the normalized baseband frequency range $\varpi \in [-\pi/2, \pi/2]$. Like any periodically sampled process, the FIR's spectrum is periodically extended in both the positive and negative frequency directions on centers that are integer multiples of the sample

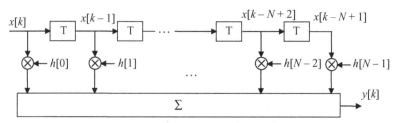

Figure 6.1 Nth-order FIR consisting of shift registers, multipliers, and accumulator.

frequency. That is, if the frequency response of a FIR, having and transfer function $H[z]$ is evaluated at $\omega = \varpi + 2\pi k$, then the resulting filter's frequency response is

$$H(e^{j\varpi+2\pi k}) = \sum_{k=0}^{N-1} h_k z^{-k} \Big|_{z=e^{j\varpi+2\pi k}} = \sum_{k=0}^{N-1} h_k e^{-jk\varpi}\left(e^{-j2\pi k}\right) = \sum_{k=0}^{N-1} h_k e^{-jk\varpi} = H(e^{j\varpi}), \quad (6.3)$$

which translates the baseband spectrum to locations defined by multiples of the normalized sample frequency ($\varpi = 2\pi$). To illustrate, a simple 10th-order FIR called a moving average (MA) filter has an impulse response given by $h[k] = (0.1)$ $[1,1,1,1,1,1,1,1,1,1]$. The filter's frequency response is $H(e^{j\varpi}) = (0.1)\Sigma(e^{jk\varpi})$, $k = 0$, $1, \ldots, 9$. The frequency response evaluated over the range $\varpi \in [-4\pi, 4\pi]$ is shown in Figure 6.2. The periodic extension of the baseband response is apparent.

IDEAL LOW-PASS FIR

An ideal low-pass filter has unit gain across a prescribed passband and infinite attenuation elsewhere. Such a filter is sometimes called a "boxcar" or "brick wall" filter and is motivated in Figure 6.3. Mathematically, an ideal low-pass filter has a frequency response given by

$$\left|H_{\text{ideal}}\left(e^{j\varpi}\right)\right| = \begin{cases} 1; & -\varpi_c \leq \varpi \leq \varpi_c \\ 0; & \text{otherwise} \end{cases}, \quad (6.4)$$

where ϖ_c is the normalized baseband cutoff frequency $\varpi_c = \omega_c/\omega_s$. From Fourier theory, the duality theorem states that the ideal low-pass filter has an impulse response given by

$$h_{\text{ideal}}[k] = \frac{\sin(\varpi_c k)}{\pi k} = \frac{\varpi_c}{\pi}\frac{\sin(\varpi_c k)}{\varpi_c k} = \frac{\varpi_c}{\pi}\text{sinc}(\varpi_c k); \; -\infty < k < \infty. \quad (6.5)$$

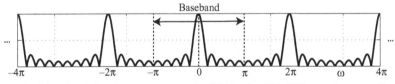

Figure 6.2   Magnitude frequency response of a 10th-order MA FIR over the normalized frequency range $\varpi \in [-4\pi, 4\pi]$.

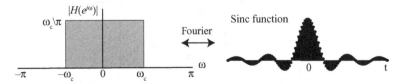

Figure 6.3   Ideal low-pass filter magnitude frequency response (left) and impulse response (right).

The impulse response is seen to have a sin(x)/x or sinc(x) shape defined for all positive and negative time. As such, the ideal low-pass filter is noncausal and is therefore not physically realizable but can, nevertheless, serve as a mathematical standard to which other low-pass filters can be compared.

Example: FIR

Consider the simple 10th-order FIR having a decaying real exponential impulse response given by $h[k] = (0.75)^k$, $k \in [0,9]$. The transfer function $H(z)$ immediately follows and it is $H(z) = \Sigma(0.75)^k z^{-k}$, $k = 0, 1, \ldots, 9$. The filter's impulse and magnitude frequency response is shown in Figure 6.4 for $\varpi \in [-\pi, \pi]$. The filter's frequency response can be seen to be decidedly low-pass and the phase response is nonlinear. The filter's magnitude frequency response, however, lacks the sharpness needed to replace an ideal sinc filter.

FIR DESIGN

The FIR design process is motivated in Figure 6.5. It begins with a set of specifications that can be articulated in a variety of forms and domains. The most common form is as a set of frequency domain specifications as summarized in Figure 6.6.

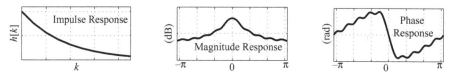

Figure 6.4   Impulse response $h[k]$ (left) two-sided magnitude frequency response (middle), and two-sided phase response (right).

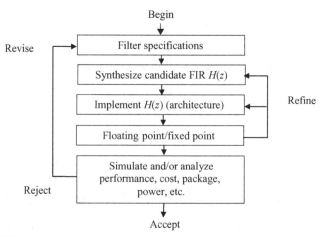

Figure 6.5   FIR design and implementation cycle.

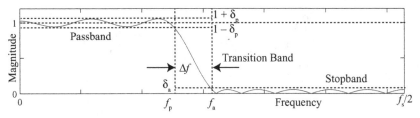

Figure 6.6   Typical FIR specifications.

Other requirements may be imposed on the final solution, such as maximum filter order, impulse response symmetry, and so forth. The design process begins by translating the solutions specifications into a filter model $H(z)$. This process is well supported with software tools that can rapidly translate the design requirement into a candidate solution. The first step, using software-enabled filter synthesis tools, is generally the easiest design step in the process. The next action involves making architecture choices. This step is particularly important because architecture strongly determines the performance envelope of an FIR. The solution can also be constructed as a fixed- or floating-point solution. Fixed-point designs are by far the most popular, if implemented in hardware. The fixed-point popularity is gained by a speed, cost, power, and memory size advantage over floating-point alternatives. If the design process breaks down at any level, corrective measures must be taken. Finally, if the end-to-end design fails to meet specifications after mitigation efforts have taken place, then a solution may require that the original set of specifications be relaxed and the design process repeated.

While specifications are assumed to define the performance envelope of a physically realizable filter, they are often framed in terms of ideal physically unrealizable filter models. To illustrate, consider that the objective is to design an approximate ideal low-pass filter. Specific side constraints might require that the realized magnitude frequency response be close to the ideal value (Fig. 6.7). The challenge becomes one of achieving the "best" solution from a list of many possible solutions. This is where filter design becomes both a science and an art.

STABILITY

If the input presented to an FIR is bounded on a sample-by-sample basis by $|x[k]| < M_x$, and the impulse response is bounded on a sample-by-sample basis by $|h[k]| < M_h$, then each partial product $h[m]x[k - m]$ of the linear convolution formula (Eq. 6.1) must be bounded by $|h[m]x[k - m]| < M_x M_h$. It then follows that

$$|y[k]| = \left| \sum_{m=0}^{N-1} h[m]x[k-m] \right| \leq \left(\sum_{m=0}^{\infty} |h[m]||x[k-m]| \right) < NM_x M_h \leq \infty. \qquad (6.6)$$

Such a filter is said to be bounded-input bounded-output (BIBO) stable. The conclusion is quite satisfying in that all FIRs with bounded coefficients are automatically

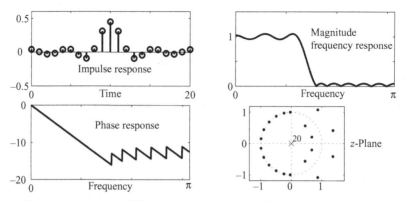

Figure 6.7 Typical low-pass FIR designed to meet a set of frequency domain constraints. Impulse response of a 21st-order FIR, (top left), FIR's magnitude frequency response (top right), phase response in radians (bottom left), and zero distribution (bottom right). The zeros off the unit circle correspond to passband locations, and those on the unit circle produce nulls and are in the filter's stopband.

stable. However, with digital filters, the issue is rarely stability, but rather quantifying the filter's run-time dynamic range requirements. This information is essential to the successful operation of a fixed-point FIR in order to insure that filters will not encounter dynamic range (register) overflow during run time. Overflow occurs whenever the data being sent to the filter's register exceed the capacity (dynamic range) of that register. The consequence of run-time overflow is the introduction of excessively large errors in the filter's output. A standard corrective action is to insure a filter, operating under worst case conditions, operates in an overflow-free mode. Specifically, a filter's worst case output is the maximum response to a worst case input. If an Nth-order FIR filter has an impulse response $h[k]$, and the filter's input $x[k]$ is bounded by M on a sample-by-sample basis, then the filter's worst case input is mathematically given by

$$x_{wc}[m-L] = M \operatorname{sign}(h[N-1-m]) = \pm M, m \in [0, N-1], \qquad (6.7)$$

where L is an arbitrary integer delay in samples. This production rule insures that each partial product in the linear convolution sum of products obtains its individual maximal positive value. Assuming $L = 0$, the worst case output to a worst case input is

$$y_{wc}[k] \le \sum_{m=0}^{N-1} h[k-m]M \operatorname{sign}(h[k=m]) = M \sum_{m=0}^{N-1} |h[k-m]| \le MG_{max}. \qquad (6.8)$$

The parameter G_{max} is called the filter's worst case gain. The worst case gain is seen to be easily computed and can serve to establish a maximal run-time dynamic range bound on the FIR's internal and external data registers and data paths. G_{max} establishes what is called the register "headroom" requirement for a fixed-point FIR. This headroom corresponds to the required dynamic range expansion of a data register in order to insure that the register does not overflow during run time. If $G_{max} \le 2^I$,

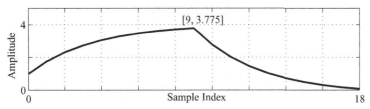

Figure 6.8   FIR's response to a worst case input.

then the corresponding FIR registers must have at least "T" additional bits of integer precision.

Example: Worst Case Analysis

Consider a 10th-order FIR having an impulse response $h[k] = (0.75)^k u[k]$. Suppose the input is a unit-bounded step function ($|u[k]| \le 1$ or $M = 1$); then, the worst case gain is given by

$$G_{\max} = \sum_{k=0}^{9} |h(k)| = 3.7747.$$

The worst case input ($L = 0$ for convenience) is computed to be $x_{wc}[m] = $ sign($h[9 - m]$) = 1 for $m \in [0, 9]$. The worst case input is a time series that is defined by the sign of the filter's impulse response in reverse time order. The response of convolving the worst case input with FIR is shown in Figure 6.8. Observe that the linear convolution sum produces a response of length $2N - 1 = 19$ samples, which is consistent with the theory of linear convolution. It can also be noted that if the input is bounded by unity, the maximal output value is $3.7747 < 2^2$, which would require at least two additional integer bits be added to the output accumulator to inhibit run-time dynamic range overflow during run time. Such an accumulator is said to be an extended precision accumulator.

LINEAR PHASE

A primary FIR attribute is its ability to precisely manage the system's phase response. This is a direct consequence of the fact that an FIR is essentially a taped delay line, which serves as an accurate phase shifter. An FIR filter is said to have a linear phase response if the measured phase response has the linear form

$$\phi(\varpi) = \alpha\varpi + \beta \tag{6.9}$$

over the normalized baseband range $\varpi \in [-\pi, \pi]$. Linear phase filters have an important role to play in a number of applications areas that are intolerant of a frequency-dependent propagation delay. Examples are

- phase lock loop (PLL) systems that are used to synchronize data and decode phase modulated signals,

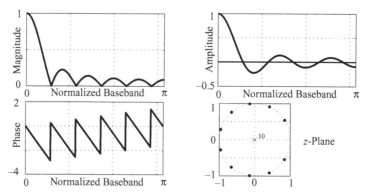

Figure 6.9   Magnitude frequency response of an 11th-order linear phase moving average (MA) FIR (top left), amplitude response (top right), phase response (bottom left), and zero distribution (bottom right).

- linear phase anti-aliasing filters placed in front of complex signal analysis subsystems (e.g., discrete Fourier transform [DFT]), and
- processing phase-sensitive signals such as images.

The frequency domain representation of an Nth-order linear phase FIR, in amplitude-phase form, is given by

$$H\left(e^{j\varpi}\right) = \sum_{k=0}^{N-1} h[k]e^{-jk\varpi} = H^{\mathrm{R}}\left(e^{j\varpi}\right)e^{-j\phi(\varpi)}, \tag{6.10}$$

where $H^{\mathrm{R}}(e^{j\varpi})$ is a real amplitude function satisfying $H^{\mathrm{R}}(e^{j\varpi}) = \pm|H(e^{j\varpi})|$. Assume that the FIR under study is a linear phase filter, where $\phi(\varpi) = \alpha\varpi + \beta$. To illustrate, Figure 6.9 displays the machine production of the filter's magnitude frequency, amplitude response, phase response, and zero locations of an 11th MA FIR having an impulse response $h[k] = (1/11)[1, 1, 1, 1, 1, 1, 1, 1, 1, 1, 1]$. The amplitude response is seen to have zero crossings and gains that alternate in sign. The phase response appears to be confused but does suggest that it may have a linear form $\theta(\varpi) = \alpha\varpi + \beta$ if properly interpreted.

Refer to the zero distribution plotted in Figure 6.9. The filter's null frequencies occur where the zeros are found on the periphery of the unit circle at the positive baseband frequency locations $\varpi = \{2\pi/11, 4\pi/11, 6\pi/11, 8\pi/11, 10\pi/11\}$. Referring to the phase response, also shown in Figure 6.9, it can be seen that the phase response appears to linear out the first null frequency $\varpi = 2\pi/11$. The phase response measured at $\varpi = 2\pi/11$ is $\phi = -10\pi/11$. The phase response can be experimentally determined by first observing that the phase y-axis intercept at $\varpi = 0$ is zero. This means that $\beta = 0$ and $\phi(\varpi) = \alpha\varpi + \beta$ (linear phase model). The phase slope is given by $\alpha = \Delta y/\Delta x = (-10\pi/11)/(2\pi/11) = -5$.

Connecting the disjoint phase segments together using a process called phase unwrapping is shown in Figure 6.10. The phase unwrapping process shown in Figure 6.10 is based on data presented in Figure 6.9. The first linear line segment of the

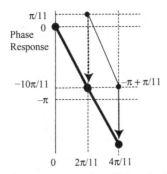

Figure 6.10   Analysis of the phase response of an 11th-order FIR showing the conversion from a piecewise linear phase spectrum into a linear phase response (using unwrapping).

phase response ranges over $\phi \in [0, -10\pi/11]$ across the normalized frequency range $\varpi \in [0, 2\pi/11]$. The next linear section ranges over $\phi \in [1\pi/11, -9\pi/11]$ across the normalized frequency range $\varpi \in [2\pi/11, 4\pi/11]$. If the second linear section is "pushed down" by π radians and attached to the end of first section, then a linear line segment is created over $[0, 4\pi/11]$. Continuing this process across the entire baseband would result in the phase response being reduced to a straight line satisfying the linear equation $\phi(\varpi) = -5\varpi$.

The previous analysis motivates the claim that a linear phase FIR can be achieved when the tap-weight coefficients are symmetrically distributed about the filter's midpoint, or center-tap coefficient. To validate this claim, consider the case presented in Equation 6.2 in a slightly modified form. Initially, it is assumed that N is odd and $L = (N - 1)/2$. Originally, the causal $H(z)$ had a point of symmetry of $k = L$. When the filter's impulse response is left shifted by L samples, the point of symmetry moves to $k = 0$ resulting in

$$H(z) = z^{-L} \sum_{m=-L}^{L} h_m z^{-m} = z^{-L} \left(h_0 + \sum_{m=1}^{L} \left(h_m z^{-m} \pm h_{-m} z^m \right) \right). \qquad (6.11)$$

The coefficients of the noncausal filter are symmetrically distributed about the center-tap coefficient h_0 with $h_i = h_{-i}$ for even symmetry and $h_i = -h_{-i}$ for odd symmetry. If N is even, then the coefficient h_0 found in Equation 6.11 is missing. The filter's frequency response can be computed by substituting $z = e^{j\varpi}$ into Equation 6.11, producing the following frequency response (shown for N odd, even symmetry):

$$H(e^{j\varpi}) = e^{-jL\varpi} \left(h_0 + \sum_{m=1}^{L} \left(h_m e^{-jm\varpi} \pm h_{-m} e^{jm\varpi} \right) \right)$$

$$= e^{-jL\varpi} \left(h_0 + \sum_{m=1}^{L} 2h_m \cos(\varpi m) \right) = e^{-jL\varpi} H^R(\varpi), \qquad (6.12)$$

where $H^R(\varpi)$ is a real amplitude function of ϖ. The FIR's phase response, defined by Equation 6.12, is $\phi(\varpi) = -L\varpi$. The claim is that such linear phase FIRs have a

TABLE 6.1. Linear Phase $\phi(\varpi) = \alpha\varpi + \beta$ FIR Relationships

Type	N (order)	Symmetry	α	β	$H_R(\varpi)$; $\varpi = 0$	$H_R(\varpi)$; $\varpi = f_s/2$
1	N odd	Even	$\alpha = -M; M = (N-1)/2$	$\beta = 0$	Unrestricted	Unrestricted
2	N even	Even	$\alpha = -M; M = (N-1)/2$	$\beta = 0$	Unrestricted	0
3	N odd	Odd	$\alpha = -M; M = (N-1)/2$	$\beta = \pm\pi/2$	0	0
4	N even	Odd	$\alpha = -M; M = (N-1)/2$	$\beta = \pm\pi/2$	0	Unrestricted

symmetric impulse response. There are, in fact, four types of symmetry that need to be discussed. They are

- N odd, even coefficient symmetry ($h_i = h_{-i}$);
- N odd, odd coefficient symmetry (antisymmetry $h_i = -h_{-i}$);
- N even, even coefficient symmetry; and
- N even, odd coefficient symmetry (antisymmetry).

The linear phase behavior of even and odd order FIRs, having even or odd coefficient symmetry, is summarized in Table 6.1 and organized as FIR Types (1–4).

The data found in Table 6.1 apply to an Nth-order FIR having a symmetric and antisymmetric impulse response. The sign of β in the linear equation $\phi(\varpi) = \alpha\varpi + \beta$ is determined by whether the antisymmetric impulse response starts positive or negative. The frequency response for each type of FIR is given by

$$H\left(e^{j\varpi}\right) = Q(\varpi)H^R(\varpi), \tag{6.13}$$

where $H^R(\varpi)$ is a zero-phase real function of ϖ and $Q(\varpi)$ is a phase-dependent term. The individual FIR filter types satisfy the following equations, for $L = (N-1)/2$:

$$\text{Type 1: } H\left(e^{j\varpi}\right) = e^{-jL\varpi}H^R(\varpi); \quad H^R(\varpi) = h[L] + 2\sum_{n=1}^{L} h[L+n]\cos(\varpi n), \tag{6.14}$$

$$\text{Type 2: } H\left(e^{j\varpi}\right) = e^{-jL_0\varpi}H^R(\varpi); \quad H^R(\varpi) = 2\sum_{n=1}^{L+1/2} h[L-1/2+n]\cos(\varpi(n-1/2)), \tag{6.15}$$

$$\text{Type 3: } H\left(e^{j\varpi}\right) = je^{-jL\varpi}H^R(\varpi); \quad H^R(\varpi) = 2\sum_{n=1}^{N_0/2} h[L+n]\sin(\varpi n), \tag{6.16}$$

$$\text{Type 4: } H\left(e^{j\varpi}\right) = je^{-jL\varpi}H^R(\varpi); \quad H^R(\varpi) = 2\sum_{n=1}^{L+1/2} h[L-1/2+n]\sin(\varpi(n-1/2)). \tag{6.17}$$

These relationships provide the foundation upon which several important classes of FIR filters are constructed.

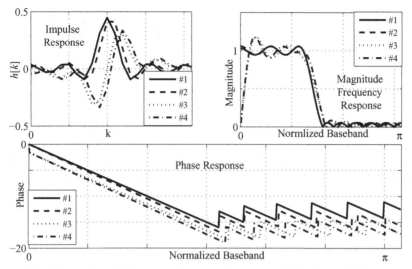

Figure 6.11 Orders 21 and 22 symmetric and antisymmetric FIRs (Types 1–4) outcomes. Impulse responses (upper left), magnitude frequency response (upper right), and phase response in radians (bottom). Notice that Types 3 and 4 filters have zero DC gain and that the phase slopes are either −10 or −10.5.

Example: FIR Types

There are four types of linear phase low-pass FIRs based on the filter order and impulse response symmetry. The four responses are illustrated below. The designed filters are

#1: $N = 21$ (odd), symmetry even, Type 1;

#2: $N = 22$ (even), symmetry even, Type 2;

#3: $N = 21$ (odd), symmetry odd, Type 3; and

#4: $N = 22$ (even), symmetry odd, Type 4.

The four design outcomes are shown in Figure 6.11. The even and odd impulse response symmetry is clearly visible. Viewing the magnitude frequency response, it can be noticed that antisymmetric FIRs have a 0 Hz (DC) gain of zero, a characteristic of this class of filter. The unwrapped phase responses are all linear with the even symmetry filters having a zero-phase DC intercept and the antisymmetric filters are displaced by $-\pi/2$. All the slopes are equal to $\alpha = -(N-1)/2$.

GROUP DELAY

It has been established that interpreting a digitally computed phase response can, at times, be challenging. A tool commonly used to quantify and/or interpret phase behavior of a filter is called group delay. Group delay is formally defined to be

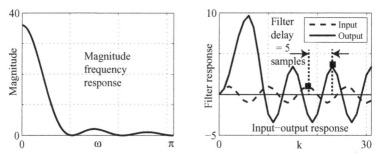

Figure 6.12 Magnitude frequency response of an 11th-order linear phase FIR (left) and response to a sinusoid at frequency $f = f_s/8$ (right). The measured propagation delay equals the group delay.

$$\tau_g = -\frac{d(\phi(\varpi))}{d\varpi} \qquad (6.18)$$

and is measured in samples. For a linear phase filter having a phase response $\phi(\varpi) = \alpha\varpi + \beta$, $\alpha = -(N-1)/2$, the group delay is

$$\tau = -\frac{d(\phi(\varpi))}{d\varpi} = -\frac{d(\alpha\varpi + \beta)}{d\varpi} = -\alpha = \frac{N-1}{2} \qquad (6.19)$$

samples for all FIR types. To illustrate, suppose an 11th-order Type 1 linear phase FIR has a symmetric impulse response $h[k] = [1, 2, 3, 4, 5, 6, 5, 4, 3, 2, 1]$. The filter has a group delay value of τ_g = five samples, which can be verified using direct computation. According to theory, the propagation delay imparted to a sinusoidal input at $f = f_s/8$ should be five samples. This can also be experimentally verified as shown in Figure 6.12.

To examine the significance of frequency-independent propagation delay, consider a 63rd-order (Type 1) linear phase FIR having a theoretical group delay $\tau_g = 31$ samples. The implication is that a signal's propagation delay through the filter is $\tau_g = 31$ sample periods or $31 \times T_s = 31/f_s$ seconds. This prediction can be experimentally investigated using sinusoidal test signals of frequencies $f = 0.1f_s$ and $0.2f_s$. An analysis of the FIR's propagation delay is shown in Figure 6.13. First consider a sinusoidal input $x[k] = \cos(2\pi k/10)$ (i.e., $f = 0.1f_s$), a signal with period $T_0 = 10T_s$. This signal appears at the filter's output as $y[k] = G_1 \cos(2\pi k/10 + \phi)$, where $G_1 = |H(e^{j2\pi/10})|$. The phase shift ϕ can be expressed in terms of the signal's period using the proportional relationship $T_0/\tau_g = (10)T_s/31T_s = 10/31$, which establishes a proportional relationship $10/31 \propto 2\pi/\phi$, or a phase shift of $\phi = 6.2\pi$. The steady-state response of the FIR can therefore be expressed as $y[k] = G_1 \cos((2\pi k/10) - 6.2\pi)$, which corresponds to a 31-sample phase shift. A similar analysis is performed for an input $x[k] = \cos(2\pi k/5)$ (i.e., $f = 0.2f_s$), and the outcome again is 31-sample delayed. Finally, the propagation delay can be directly measured from the data presented in Figure 6.13. A point on the input signal is seen to appear at the output 31 samples later for both signals. The conclusion is that in both test cases, the propagation delay is 31 samples regardless of the input frequency.

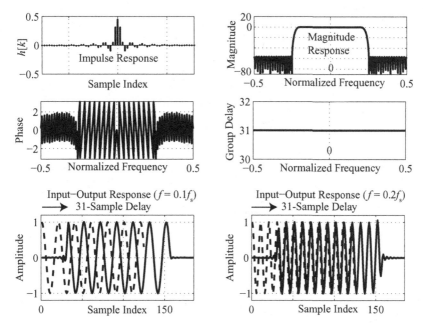

Figure 6.13   FIR impulse response and magnitude frequency response (top); phase and group delay (middle). Input–output response for sinusoidal inputs at frequencies $f = 0.1f_s$ and $0.2f_s$ sinusoidal input showing a 31-sample delay relationship (bottom).

FIR ZERO LOCATIONS

The zeros of a linear phase Nth-order FIR, having a transfer function $H(z)$, satisfies the algebraic relationship

$$z^N H(z) = H^*(1/z^*). \qquad (6.20)$$

That is, if z is a zero of $H(z)$, then $(1/z^*)$ must also be a zero of $H(z)$. This fact translates into the following possible cases:

- Zeros can be located at $z_i = \pm 1$ since $(1/z_i^*) = \pm 1$.
- Zeros can be located on the unit circle at $z = z_i$ and $z = z_i^*$ since $(z - z_i)$ $(z - z_i^*) = (z - (1/z_i^*))(z - (1/z_i))$.
- The real zeros lying off the unit circle must occur in reciprocal pairs.
- The complex zeros lying off the unit circle must occur in groups of four, $\{z_i, z_i^*\}$, and their reciprocals.

These conditions are graphically interpreted in Figure 6.14.

Example: Zero Distribution

A linear phase Type 1 21st-order low-pass FIR has the zero distribution shown in Figure 6.15. The zeros located on the unit circle (stopband) appear with

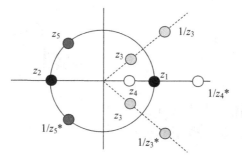

Figure 6.14 Possible zero distribution of a linear phase FIR.

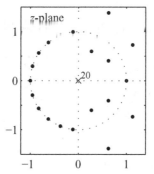

Figure 6.15 Zero distribution of a 21st-order linear phase Type 1 FIR.

complex-conjugate symmetry. Those zeros off the unit circle (passband) appear with complex-conjugate and reciprocal symmetry, as shown in Figure 6.14.

ZERO-PHASE FIR

A zero-phase filter is a special case of a linear phase filter having a phase response $\phi(\varpi) = \alpha\varpi + \beta = 0$. First, the FIR must have even symmetry in order to insure that $\beta = 0$. If $\alpha = 0$, then group delay must likewise be zero, suggesting that the filter has a zero delay. Suppose an Nth-order Type 1 linear phase FIR, having an impulse response $h[k] = \{h[0], h[1], \ldots, h[N-1]\}$ and coefficient symmetry has a center tap located at $L = (N-1)/2$. For the Type 1 FIR, the filter's frequency response is given by

$$H(e^{j\varpi}) = H^R(e^{j\varpi})e^{-j\tau_g\varpi}, \qquad (6.21)$$

where τ_g is the group delay and $H^R(e^{j\varpi})$ is a real amplitude. The filter described by the impulse response is causal, having its initial sample $h[0]$. If the center-tap coefficient (point of symmetry) is left-shifted by τ_g samples, then a new noncausal filter results, having an impulse response $h_0[k] = \{h[-L], h[-(L-1)], \ldots, h[-1], h[0], h[1], \ldots, h[L-1], h[L]\}$ where $L = (N-1)/2 = -\tau_g$. The resulting FIR's frequency response is

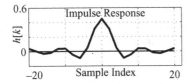 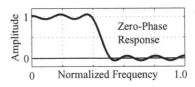

Figure 6.16   Zero-phase FIR (left), real frequency response (right).

Figure 6.17   Zero-phase architecture.

$$H_0(e^{j\omega}) = \left(H^R(e^{j\omega}) e^{-j\tau_g \omega} \right) e^{j\tau_g \omega} = H^R(e^{j\omega}), \qquad (6.22)$$

where $H_0(e^{j\omega})$ and $H^R(e^{j\omega})$ are real, and is called a zero-phase filter. The zero-phase FIR is seen to be noncausal and therefore cannot be used in real-time applications. The filter can, however, be used in non-real-time or off-line applications requiring zero delay filtering. This is commonplace in such areas such as image processing.

Example: Zero-Phase FIR

A 21st-order linear phase FIR is converted into a zero-phase filter by left shifting the original impulse response by $\tau_g = 10$ samples. The resulting zero-phase FIR impulse response and real frequency response $H^R(e^{j\omega})$ is shown in Figure 6.16 and indicates both positive and negative frequency-dependent gain.

   A zero-phase filter can also be synthesized as a quadratic filter function $|H(z)|^2$, where

$$|H(z)|^2 = H(z)H(z^{-1}) = H_{zp}(z) \Leftrightarrow h[k] \otimes h[-k] = h_{zp}[k], \qquad (6.23)$$

where $H(z)$ is an Nth-order FIR. The design strategy is motivated in Figure 6.17. Since $|H(z)|^2$ is entirely real, the resulting systems is also called a zero-phase filter. Referring to Equation 6.23, a zero-phase filter is seen to be of order $(2N - 1)$, which is always odd. The point of coefficient symmetry is also located at $k = 0$.

MINIMUM PHASE FILTERS

An FIR having all zeros interior to the unit circle is called a minimum phase filter. Minimum phase FIR systems have a nonlinear phase response. Compared with a linear phase FIR and a given magnitude frequency response specification, a minimum phase realization normally provides a lower order solution.

   Minimum phase filters can be thought of as "minimum delay" filters in that they have less propagation delay than linear phase filters, while possessing essentially the same magnitude frequency response. While a low-pass linear phase FIR filter has its largest (magnitude) coefficients in the center of the impulse response, the largest (magnitude) coefficients of a minimum phase filter are found at the beginning. Minimum phase FIRs can be synthesized from real symmetric linear phase FIRs by reflecting the filter's zeros residing outside the unit circle, to interior locations. Recall that the zeros of a linear phase FIR naturally group themselves into pairs of conjugate pairs of the form $(\alpha + j\beta)$, $(\alpha - j\beta)$, $1/(\alpha + j\beta)$, $1/(\alpha - j\beta)$. By reflecting the exterior zero to a reciprocal interior point, the FIR's zeros become $(\alpha + j\beta)$, $(\alpha - j\beta)$, $(\alpha + j\beta)$, $(\alpha - j\beta)$, or a doubling of the zeros at an interior position without compromising the magnitude frequency response. The problem with this scheme is that the resulting minimum phase FIR has the order of the linear phase FIR. Another method, called spectral factorization, is an extension of the reflection technique. This method is based on the idea that $|H(z)|^2 = H(z)H^*(z^{-1})$, where the zeros are redistributed to $H(z)$ and $H^*(z^{-1})$ so as to cause $H(z)$ to be a minimum phase FIR. The multistep design procedure is shown below:

- Design a prototype FIR that satisfies $|H(z)|^2$ specifications.
- Compute the zeros of the prototype FIR.
- Delete all zeros residing outside the unit circle and ½ of all the zeros residing on the unit circle.
- Convert the surviving zeros into the coefficients of an FIR and gain-adjust the new FIR.

Some refinements are needed to insure that the prototype filter correlates to the specifications of $H(z)$ and retains minimum phase behavior. If the desired passband ripple deviation δ_p and stopband ripple deviation δ_a are given, then the prototype passband ripple deviation should be assigned a value

$$\delta_{p(\text{prototype})} = \frac{4\delta_p}{\left(2 + 2\delta_p^2 - \delta_a^2\right)} \tag{6.24}$$

and stopband value targeted to be

$$\delta_{a(\text{prototype})} = \frac{\delta_a^2}{\left(2 + 2\delta_p^2 - \delta_a^2\right)}. \tag{6.25}$$

The prototype requirements can lead to a higher-order filter, twice the order of the final minimum phase filter $H(z)$. A number of MATLAB objects support the design of minimum phase filters. Included in this list are the firminphase and firgr (equiripple) programs.

Example: Minimum Phase FIRs

A 21st linear phase and minimum phase FIR is designed to meet or exceed the previous set of magnitude frequency response constraints. The distribution of the FIR's zeros is displayed in Figure 6.18. The minimum phase zeros are seen to be on or

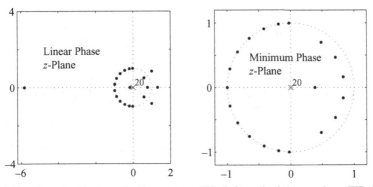

Figure 6.18   Zero distribution of a linear phase FIR (left) and minimum phase FIR (right).

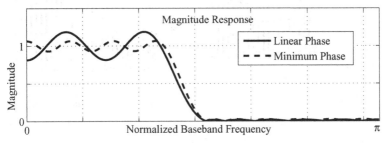

Figure 6.19   Magnitude frequency response of the linear phase and minimum phase FIRs.

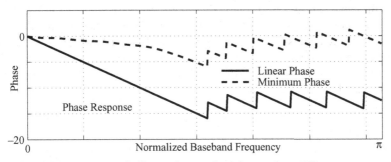

Figure 6.20   Phase response of a linear phase and minimum phase FIR.

interior to the unit circle. The magnitude frequency responses of both FIR filters are shown in Figure 6.19. While the linear phase FIR meets the specifications, the minimum phase FIR significantly exceeds them. This suggests that a minimum phase FIR of order less than 21 could be designed that meets the posted specifications. The phase responses of the two FIRs are displayed in Figure 6.20. Over the passband, the minimum phase FIR is seen to have a nonlinear phase response.

The impulse response of each FIR is shown in Figure 6.21. The minimum phase FIR is seen to produce a strong signal early in the impulse response's time history. The linear phase FIR, however, concentrates its activity to a much later time,

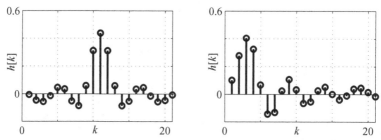

Figure 6.21 Impulse responses of linear phase FIR (left) and minimum phase FIR (right).

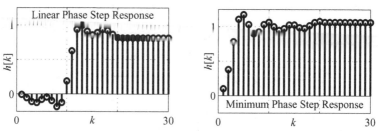

Figure 6.22 Step response of a linear phase (left) and a minimum phase FIR (right).

in fact around the τ_g sample. One of the features that can make a minimum phase filter useful is the rapid development of the impulse response as motivated in Figure 6.21. This can result in a more aggressive forced response as shown in Figure 6.22, which shows the step response of both filters. It can be seen that the minimum phase response is more aggressive than that of a linear phase FIR. While the steady-state behavior of both systems is the same, the minimum phase filter achieves steady-state status earlier. This can be an important consideration when choosing a filter for use in control applications.

CHAPTER 7

WINDOW DESIGN METHOD

FINITE IMPULSE RESPONSE (FIR) SYNTHESIS

Normally, FIRs are designed to meet a set of frequency domain specifications defined over the frequency range $\varpi \in [-\pi, \pi]$, where $\varpi = \pi$ corresponds to the normalized Nyquist frequency. Specifically, it is assumed that the target FIR has a frequency response given by $H_d(e^{j\varpi}) = |H_d(e^{j\varpi})|\angle H_d(e^{j\varpi})$. The objective of any FIR design strategy is to synthesize a physically realizable filter having a frequency response $H(e^{j\varpi})$ that closely approximates the desired response $H_d(e^{j\varpi})$ in some acceptable manner.

The metrics of comparison are generally defined in terms of an error criterion where, in the frequency domain, the approximation error is given by

$$\varepsilon(\varpi) = \left| H_d\left(e^{j\varpi}\right) - H\left(e^{j\varpi}\right) \right|. \tag{7.1}$$

A popular FIR design strategy that attempts to minimize the approximation error is called minimum squared error (MSE) criterion and is given by

$$\text{minimize } \phi(\varpi) = \sum_{\forall \varpi} e(\varpi)^2 ; \forall \varpi \in [-\pi, \pi) \tag{7.2}$$

over the normalized baseband frequency range as illustrated in Figure 7.1. The advantage of the MSE method is that it is well known. The shortcoming of the MSE method is found in the fact that even though the majority of baseband frequency errors can be small, large localized errors can occur. Common MSE FIR design strategies have been reduced to software; therefore, creating these filters and evaluating them is no longer an obstacle.

WINDOW-BASED DESIGN

One of the simplest FIR design methods begins with an ideal (a.k.a., boxcar) low-pass filter model having a noncausal impulse response:

$$h[k] = \frac{\sin(\pi k / 2)}{\pi k}. \tag{7.3}$$

Digital Filters: Principles and Applications with MATLAB, First Edition. Fred J. Taylor.
© 2012 by the Institute of Electrical and Electronics Engineers, Inc.
Published 2012 by John Wiley & Sons, Inc.

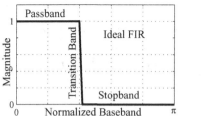 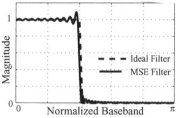

Figure 7.1 Example of a typical FIR design criteria (left). Ideal and MSE filter response (right). Note the proportionally large localized error found at the transition band edge.

The filter described in Equation 7.3 is infinitely long and therefore not physically realizable. However, if a suitable finite duration approximation of the ideal impulse response shown in Equation 7.3 could be found, then a physically realizable approximate solution is possible. A finite N-sample approximating impulse response can be extracted from the ideal impulse response using an N-sample rectangular window, or mask, denoted $w_N[k]$, where

$$w_N[k] = \begin{cases} 1; & |k| \le (N-1)/2 \\ 0: & \text{otherwise} \end{cases}, \tag{7.4}$$

which can be used to define a finite duration time series $h_N[k] = w[k] \times h[k]$. The fact that $h_N[k]$ extends into negative time means that the windowed filter $h_N[k]$ is noncausal and therefore unable to support real-time applications. In order to convert the finite noncausal FIR into a causal FIR, the windowed impulse response would need to be right-shifted in time by an amount $L = (N-1)/2$ samples. The resulting causal impulse response is given by

$$h_C[k] = \frac{\sin((\pi/2)(k-(N-1)/2))}{\pi(k-(N-1)/2)}; k = 0, 1, 2, \dots, (N-1). \tag{7.5}$$

The design process is motivated in Figure 7.2. Because the windowing (masking) operation is applied to $h[k]$, the design strategy is called window-based or window method. Referring to Figure 7.2, it can be seen that the magnitude frequency response of the FIR filter, defined by Equation 7.5, exhibits an oscillatory error with the largest errors being located about the points of discontinuity in the desired boxcar response (i.e., transition band). This behavior is traditionally called Gibbs phenomena or Gibbs error.

There are four possible types of ideal piecewise-constant filter models that need addressing, and they are shown in Figure 7.3. The production rules used to produce each ideal noncausal filter models are shown in Table 7.1.

DETERMINISTIC DESIGN

Using the window method, design a 201st-order linear phase low-pass (i.e., boxcar) filter having a near unit passband out to $\varpi_c = 0.1\varpi_s$, and a gain at, or near, 0 across

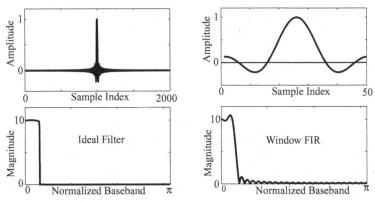

Figure 7.2   Windowing of an ideal low-pass filter's impulse response (upper left) with a 51-sample finite rectangular window $w[k]$ (upper right). Notice that the magnitude frequency response changes from being nearly ideal to one having a discernable oscillating approximation error that becomes locally large near the edges of the filter's transition band.

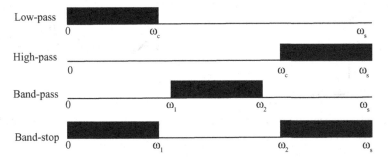

Figure 7.3   Magnitude frequency response of ideal noncausal filter models.

TABLE 7.1.   Ideal Impulse Responses (f_c, f_1, f_2 Are Cutoff Frequencies)

Type	$h[k]$	$h[0]$
Low-pass	$2f_c \sin(k\varpi_c)/k\varpi_c$	$2f_c$
High-pass	$-2f_c \sin(k\varpi_c)/k\varpi_c$	$1 - 2f_c$
Band-pass	$(2f_2 \sin(k\varpi_2)/k\varpi_2) - (2f_1 \sin(k\varpi_1)/k\varpi_1)$	$2(f_2 - f_1)$
Band-stop	$(2f_1 \sin(k\varpi_1)/k\varpi_1) - (2f_2 \sin(k\varpi_2)/k\varpi_2)$	$1 - 2(f_2 - f_1)$

the stopband. The inverse transform and therefore the FIR coefficients are mathematically given by the sample values defined by Equation 7.5, namely,

$$h[k] = \frac{\sin\left(\varpi_c \left(k - \tau_g\right)\right)}{\pi\left(k - \tau_g\right)}, \tag{7.6}$$

where τ_g is the FIR's group delay. The specific truncated $\sin(x)/x$ impulse response of the causal 201st-order low-pass FIR, having a cutoff frequency of $f_c = 0.1f_s$, is generated and displayed in Figure 7.4. The frequency response is seen to have the general shape of an ideal low-pass filter, but exhibits some local excessive Gibbs errors about the edges of the transition band.

The higher the filter order N, the better the approximation to an ideal low-pass filter model. However, Gibbs errors are persistent and typically result in a 9% error before and after a discontinuity. The key parameters of comparison include the realized transition bandwidth and passband and stopband critical frequencies. Finally, the mean squared approximation error measure is given by Equation 7.1. The approximation error can be somewhat excessive due, in part, to the fact that the rectangular window has a $\sin(x)/x$ response having locally weak stopband attenuation. Other choices of windows, such as Hamming, offer better trade-offs between passband and stopband gain. Commonly used fixed coefficient window choices and their key parameters are reported in Table 7.2 and Figure 7.5. It can be seen that each window brings to the table its unique set of performance parameters and produce different outcomes.

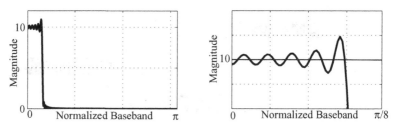

Figure 7.4   Frequency response of a 201st-order $\sin(x)/x$ impulse response (left); passband details (right).

TABLE 7.2.   Common Fixed-Coefficient N-Sample Windows

Window	Main Lobe Width	−3-dB Width	−6-dB Width	Max. Side Lobe (dB)	Side Lobe Roll-Off dB/dec.	Normalized Transition Bandwidth	Passband Ripple
Rectangle	$4\pi/N$	0.88	1.21	−13	29	0.9/N	0.74
Hamming	$8\pi/N$	1.30	1.8′	−43	20	3.3/N	0.02
Hann	$8\pi/N$	1.44	2.00	−32	60	3.1/N	0.055
Blackman	$12\pi/N$	1.64	2.3	−58	60	5.5/N	0.0017

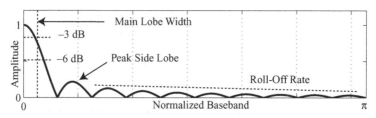

Figure 7.5   Typical frequency domain window specifications.

DATA WINDOWS

A rectangular window can introduce errors that are attributable to Gibbs phenomenon. Numerous data smoothing windows have been developed that can suppress this type of error. Some of the popular fixed coefficient windows that meet these criteria are listed in Table 7.2 (rectangular included for completeness), with a few graphically interpreted in the time and frequency domain in Figure 7.6. It should be noted that there are many more windows appearing in the literature. The windows found in Table 7.2 are simply some of the more popular representatives from the class of fixed-coefficient windows. MATLAB, for example, supports the following recognized window functions:

Barllett	Chebyshev	Hann	Rectangular
Blackman	Flat Top	Kaiser (variable)	Taylor
Blackman-Harris	Gaussian	Nuttall	Triangle
Bohman	Hamming	Parzen	Tukey

All of the windows in Table 7.2, except the rectangular window, are seen to have sample values at or near zero at the "ends" of their support interval (i.e., $k = 0$ and $k = N - 1$). This eliminates or reduces any potential jump discontinuities located at the ends of the window. Since Gibbs phenomenon is attributed to such discontinuities, windows can suppress this type of distortion.

There are also several parameterizable windows that augment the list of fixed-coefficient windows presented in the list. One popular parameterized window is the Kaiser window given by

$$w[k] = \frac{I_0\left(\beta\sqrt{1-(k/M)^2}\right)}{I_0(\beta)}; -M \le k \le M, \tag{7.7}$$

where $I_0(\beta)$ is the modified 0th-order Bessel function. The parameter β affects the main lobe bandwidth and stopband attenuation trade-offs. The time- and frequency-domain behavior of a Kaiser window is shown for $\beta = 1$, 3, and 10 in Figure 7.7. The selected value of β is influenced by the desired attenuation band gain A as illustrated below:

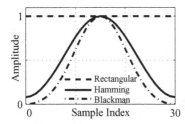 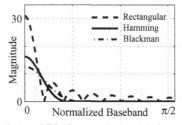

Figure 7.6 Thirty-first-order rectangular, Hamming and Blackman windows (left) and their magnitude frequency responses (right).

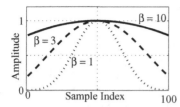 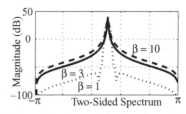

Figure 7.7   Kaiser parameterizable window with time-domain profile shown on the left and two-sided magnitude frequency response on the right.

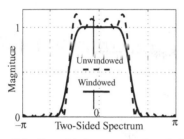 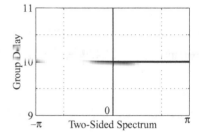

Figure 7.8   Two-sided spectrum of a windowed (rectangular and Hamming) FIR. Dual spectrum (left); windowed group delay (right).

$$\beta = 0 \text{ if } A \le -21 \text{ dB,}$$
$$\beta = 0.584(A - 21)0.4 + 0.079(A - 21) \text{ if } -21 \le A \le 50 \text{ dB,} \tag{7.8}$$
$$\beta = 0.1102(A - 8.7) \text{ if } A \ge 50 \text{ dB.}$$

Example: Window-Based FIR Design

Consider the design of a 21st-order linear phase Type I low-pass FIR design that approximates an ideal filter. The 21st-order linear phase FIR has a group delay given by $\phi(\varpi) = -\tau_g \, \varpi = -10\varpi$. If a 21-sample Hamming window is applied to the 21st-order FIR, a Hamming-windowed FIR results. The magnitude frequency responses of the rectangular and Hamming-windowed filters are shown in Figure 7.8 with the rectangular windowed version exhibiting considerable Gibbs distortion about the transition band. Observe that the Hamming window suppresses the large Gibbs error previously located near the transition band edge. This reduction in error is gained at the expense of a widened transition band (factor of 3.6x from Table 7.2). Since a Hamming window possesses even symmetry; the linear phase property and group delay of the windowed Type I FIR is unaltered.

MATLAB WINDOW FIR DESIGN

MATLAB provides a tool called fir1 that can be used to design a linear phase FIR using the window method. The program accepts standard MATLAB windows and

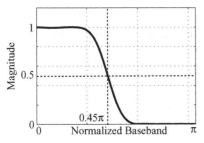 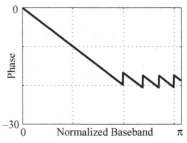

Figure 7.9   A 21st-order (20th order in MATLAB) low-pass linear phase FIR designed using *fir1* and a Hamming window. Magnitude frequency response (left) and phase response (right).

can be used to design a low-pass, high-pass, band-pass, or band-stop as well as multiband filters. Windows can be multiplicatively applied to an FIR's impulse response anytime. The MATLAB fir1 syntax is given by

```
h = fir1(n,wn,'ftype', window)
```

where *n* is the MATLAB filter order (order = *n* + 1 form FIR theory)*; "*wn*" is passband cutoff frequencies normalized to the Nyquist frequency (not the sample frequency f_s); *ftype* = *low-pass*, *high-pass*, *band-pass*, and *band-stop*; and window is the user-selected data window function (default = Hamming). MATLAB recommends the use of fir2, however, for designing FIRs having an arbitrary magnitude frequency response. To illustrate, consider designing a low-pass FIR having a critical passband frequency of $f_c = 0.45(f_s/2)$, a baseband frequency that specifies a −6-dB (i.e., 0.5) filter gain. A low-pass filter having a −6-dB gain at $f_c = 0.45(f_s/2)$ is shown in Figure 7.9.

KAISER WINDOW

Parameterizable windows, such as a Kaiser window, are more versatile than a fixed parameter window. In order to use a Kaiser window in the design of an FIR, the Kaiser window needs to be properly parameterized. A formula that predicts a reasonably accurate order estimate for a Kaiser windowed FIR is defined in terms of transition bandwidth (Δf), sample period (T_s), and passband and attenuation error deviations δ_p and δ_a. The MATLAB-computed order estimate is given by

$$N = \frac{-20\log(\delta) - 7.95}{14.36\Delta f T_s}, \tag{7.9}$$

where $\delta = \min(\delta_p, \delta_a)$. The veracity of the estimate is judged on a case-by-case basis.

---

* MATLAB caveat: MATLAB defines an FIR filter having an impulse response $h[k] = \{h[0], h[1], \dots, h[n]\}$ to be *n*, whereas the order is normally defined to be *n* + 1 in the literature and classic filter analysis.

Example: FIR Order

Suppose a windowed band-pass FIR is desired having a passband cutoff frequency of $f_c = (0.2 \pm 0.05)f_s$, transition bandwidth of $\Delta f = (0.05)f_s$, tolerable passband error or ripple of 0.2 dB, and the stopband attenuation of at least 60 dB. The passband and stopband error deviations are

$$\delta_p = 0.012 \text{ (such that } 20\log_{10}(1+\delta_p) = 0.2 \text{ dB)}$$

and

$$\delta_a = -60 \text{ dB or } \delta_a = 10^{-3},$$

then $\delta = \min(\delta_p, \delta_a) = 10^{-3}$. From these parameters, a Kaiser-windowed FIR of order

$$N = \frac{-20\log(\delta) - 7.95}{14.36\Delta f\, T_s} = \frac{60 - 7.95}{14.36(0.05)} \sim 73$$

is required. From Equation 7.8 and $A \geq 50$ dB, it follows that $\beta = 0.1102(A - 8.7) = 5.6$. The production of these parameters can also be facilitated using MATLAB and MATLAB function `kaiserord` as exemplified by executing the code shown below:

```
f=[.1 .15 .25 .3]; % critical frequencies

a=[0 1 0 ]; % gain profile - band-pass filter

[N,Wn,BTA,FILTYPE]=kaiserord(f,a,[.001 .012 .001],1);

[N = 73 order, Wn = 0.2500, 0.5500, normalized by the Nyquist frequency]

BTA =5.6533 %(beta)
```

A Blackman window solution is also possible. It follows that the order of the Blackman window is defined by the transition band bandwidth, namely, $N = 5.5/\Delta\phi = 110$. The Kaiser and Blackman designs are presented in Figure 7.10.

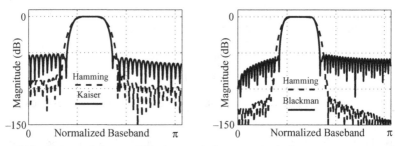

Figure 7.10   Kaiser window and Blackman window magnitude frequency responses (left) and log magnitude (right).

TRUNCATED FOURIER TRANSFORM DESIGN METHOD

Another simple FIR synthesis technique is called the truncated Fourier transform method.* It is based on the use of an inverse discrete Fourier transform, or IDFT, to produce a database that is used to define an Nth-order physically realizable FIR that emulates the shape and form to the desired FIR frequency response. This method provides a straightforward means of synthesizing an Nth-order FIR from an arbitrary desired frequency response specification. Formally the method involved computing an M-sample ($M \gg N$) IDFT of the desired filter's magnitude frequency response ($|H_d(e^{j\varpi})|$) and phase response ($\angle H_d(e^{j\varpi})$). The M-sample time series $h[k]$, resulting from an M-point IDFT, is then reduced (pruned) to an N-sample realizable FIR. For linear phase filters, the realized filter is centered about sample $L = (M/2) - 1$ and the phase is set to $\phi(\varpi) = (N - 1)\varpi/2$ for a Type 1 linear phase FIR. The N-samples symmetrically extracted about the filter's midpoint forms an Nth-order linear phase FIR whose magnitude frequency response approximates the desired response. This process is motivated in Figure 7.11.

Types 2, 3, and 4, as well as nonlinear phase response FIRs, are also possible. For a Type 1 FIR, the desired frequency response is

$$H\left(e^{j\varpi}\right) = \left|H\left(e^{j\varpi}\right)\right|\left(\cos(\phi(\varpi)) + j\sin(\phi(\varpi))\right). \tag{7.10}$$

Reviewing the method, the frequency-domain specifications of the desired filter interprets Equation 7.10 at the frequency locations $\varpi_i = i2\pi/M$, $i \in [-M/2, M/2)$ where $\varpi_i \in [-\pi, \pi)$. The long M-point IDFT outcome is then reduced to an N-point realizable FIR impulse response using symmetric truncation. It is important to note, however, that there are really no restrictions on the form and shape of the desired filter's frequency response. As long as it can be represented by an IDFT, a realizable filter can be extracted by windowing.

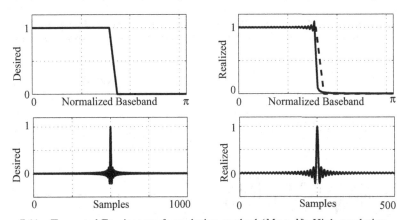

Figure 7.11   Truncated Fourier transform design method ($M \gg N$). High-resolution spectrum (upper left) and realized spectrum (upper right). High-resolution impulse response (lower left) and truncated (realized) impulse response (lower right).

---

* The truncated Fourier method is sometimes called the frequency sampling technique.

Example: Truncated Fourier Transform Design Method

To illustrate the truncated Fourier method, consider the design or a rectangular windowed 21st-order linear phase Type I FIR. The desired filter's magnitude frequency response is given by

$$|H_d(e^{j\varpi})| = 1 \text{ for } 0 \le \varpi \le 0.4\pi \text{ (passband)},$$

and

$$|H_d(e^{j\varpi})| = 1 - (10/\pi)(\varpi - 0.4\pi) \text{ for } 0.4\pi \le \varpi \le 0.5\pi \text{ (transition band)},$$
$$|H_d(e^{j\varpi})| = 0 \text{ for } 0.5\pi \le \varpi \text{ (stopband)}.$$

A desired Type 1 linear phase response for the 21st-order FIR is specified using $\phi(\varpi) = -\tau_g \varpi = -10\varpi$. The resulting filter is shown in Figure 7.12.

The desired magnitude frequency response is first mapped into the time domain using a 1023-point inverse fast Fourier transform (IFFT). The realized FIR's 21 coefficients are extracted from a 1023-point IFFT's midpoint, or at $k = 511$. The realized magnitude frequency response is seen to be a good approximation to the desired magnitude frequency response. The maximum error occurs at the boundary of the transition band and is attributable to the Gibbs phenomenon. The phase response is linear and the filter's group delay is 10 samples.

It is also possible to apply classic data windows (e.g., Hamming) to the synthesized $h[k]$ in order to mitigate Gibbs effects or excessive overshoot. While applying a data window to achieve these goals, it should be noted that data windows have a detrimental effect on transition bandwidth. As a result, the generated realizable filter must always be checked against the design specifications to insure compliance.

FREQUENCY SAMPLING DESIGN METHOD

The frequency sampling design technique can formally be used to design an Nth-order FIR based on the presumption that the filter's frequency response is known at

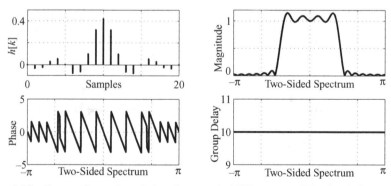

Figure 7.12 Twenty-first-order window (rectangular) FIR truncated Fourier design example showing realized impulse response (top left), two-sided spectrum (top right), phase response (lower left), and group delay (lower right).

N periodically spaced harmonics. Mathematically, a frequency sampling filter's response is defined in terms of

$$h[k] = \frac{1}{N} \sum_{n=0}^{N-1} H(n) W_N^{nk}. \tag{7.11}$$

The filter's impulse response is a function of the filter's frequency response evaluated on harmonic centers with spacing $f_0 = \pm f_s/N$. A discipline is required to insure that the resulting filter has real coefficients. There are several frequency assignment schemes that can be used to assign the harmonic values. The details of the inverted spectra can vary as a function of the frequency assignments. In the case where the FIR is to be designed using an IDFT or FFT. The frequency location $\varpi = 0$ must be present. The frequency spacing is then determined by the filter order N. A linear phase odd order FIR with even coefficient symmetry is given by

$$h[k] = \frac{1}{N} \sum_{k=1}^{(N/2)-1} 2|H(n)| \cos(2\pi n(k-L)/N) + H(0); L = \frac{N-1}{2}. \tag{7.12}$$

It can be seen that the preassigned spectral values are mapped to a real impulse response. Each even or odd order FIR, with even or odd symmetry, would be generated by a similar equation.

Example: Frequency Sampling Filter

MATLAB provides a tool, fir2, that produces an FIR filter using the frequency sampling method. The fir2 program produces a linear phase FIR with an applied default "Hamming" window. Other windows can be used, such as a rectangular window (rectwin). Use fir2 to design of a frequency sampling low-pass FIR that approximates a desired ideal (boxcar) magnitude frequency response with a cutoff frequency of $f_c = 0.45 f_N$, $f_N = f_s/2$. The fir2 program accepts an array of desired frequencies and corresponding magnitude frequency responses and converts the data into a Nth-order FIR. The program explicitly requires that the frequency/gain pair be defined for $\varpi = 0$ and the Nyquist frequency. This restricts the adoption of some of the possible mathematical frequency assignment rules, but results in an algorithm that can be based on the use of the IDFT or FFT.

The first design attempts to synthesize an ideal low-pass filter having the passband end and a stopband beginning at $f_c = 0.45 f_N$ (0 transition band), using both rectangular and Hamming windows. A second version creates a transition band having a width of $f_t = 0.1 f_N$, again using rectangular and Hamming windows. This model linearly interpolates the transition band. The third design forces a more reasonable transition band behavior on the filter by subdividing the transition band into subfrequencies $f_1 = 0.4833 f_N$ and $f_2 = 0.5167 f_N$ with gains of 0.8 and 0.2, respectively, again using rectangular and Hamming windows. The results are shown in Figure 7.13 for the rectangular and Hamming window case. With the rectangular window in use, the ideal model (0 transition bandwidth) is seen to approximately

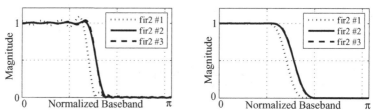

Figure 7.13   Frequency sampling filter design using rectangular window (left) showing frequency response. Frequency sampling filter design using a Hamming window (right) showing frequency response.

follow their assigned transition band trajectories. The linearly interpolated and piecewise linearly interpolated responses are seen to follow the form of the defined transition band gain profile. Figure 7.13 shows the Hamming window case. When a Hamming window is applied, a reduction in Gibbs error is traded off against a widened transition band.

CHAPTER 8

LMS DESIGN METHOD

FINITE IMPULSE RESPONSE (FIR) SYNTHESIS

Computer-generated FIR design strategies fall into several categories based on a chosen optimization criteria. In the frequency domain, the approximation error is assumed to be given by

$$\varepsilon(\varpi) = \left| H_d\left(e^{j\varpi}\right) - H\left(e^{j\varpi}\right) \right|. \tag{8.1}$$

One of the more popular and useful design criteria is one that minimized the error squared and is given by

$$\text{minimize } \phi(\varpi) = \sum_{\forall \varpi} e(\varpi)^2. \tag{8.2}$$

An example of a least mean square (LMS) approximation to an ideal boxcar filter is shown in Figure 8.1. There is a good general agreement between the LMS and boxcar responses with the possible exception of large localized errors near points of discontinuity or transition band.

LEAST-SQUARES METHOD

The least-squares method provides engineers and scientists with a robust means of approximating the solution of an overdetermined system (i.e., more equations than unknowns). Least-squares analysis produces a means of obtaining a solution that minimizes the sum of the squares produced by each individual equation. One of the most important least-squares applications is in data fitting, which includes determining what FIR "best" approximates a desired frequency response. The general least-squares method was described by Carl Friedrich Gauss. Using this classic mathematical tool, an FIR filter can be synthesized that can meet very demanding specifications.

Digital Filters: Principles and Applications with MATLAB, First Edition. Fred J. Taylor.
© 2012 by the Institute of Electrical and Electronics Engineers, Inc.
Published 2012 by John Wiley & Sons, Inc.

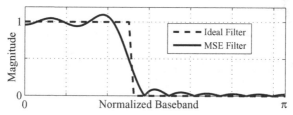

Figure 8.1   LMS FIR design strategy. MSE, mean squared error.

Gauss has made contributions to many scientific fields, including number theory, statistics, analysis, differential geometry, geodesy, geophysics, electrostatics, astronomy, and optics. Gauss is credited with developing the fundamentals that form the basis for least squares analysis in 1795 at the age of 18. Legendre, however, was the first to publish the method. Gauss realized that the least-squares method could be used to analyze an overdetermined system with measurement y_i and unknown coefficient β_i, which satisfy a linear equation:

$$\sum_{i=1}^{m} x_{ij}\beta_j = y_i; \ j \in [1, n], m > n. \tag{8.3}$$

The resulting m linear equations are defined in terms of n unknowns that can be written in linear algebraic form as

$$X = \begin{bmatrix} x_{11} & x_{12} & \ldots & x_{1n} \\ x_{21} & x_{22} & \ldots & x_{2n} \\ \ldots & \ldots & \ldots & \ldots \\ x_{m1} & x_{m2} & \ldots & x_{mn} \end{bmatrix}; \beta = \begin{bmatrix} \beta_1 \\ \beta_2 \\ \ldots \\ \beta_n \end{bmatrix}; y = \begin{bmatrix} y_1 \\ y_2 \\ \ldots \\ y_m \end{bmatrix}; X\beta = y. \tag{8.4}$$

Johann Carl Friedrich Gauss (1777–1855)

The LMS solution is one that minimizes:

$$\min_{\beta} \sum_{i=1}^{m} \left| y_i - \sum_{j=1}^{n} x_{ij}\beta_j \right|^2 . \tag{8.5}$$

The optimal LMS solution satisfies $\hat{\beta} = \left(X^T X \right)^{-1} X^T y$.

LEAST-SQUARES FIR DESIGN

Classic least-squares estimation techniques can be applied to the FIR design question. Consider designing a linear phase FIR that minimizes the weighted minimum squared error (MSE) criterion,

$$\sigma = \sum_{i=1}^{K} \left\{ W(\omega_i)[e(\varpi_i)] \right\}^2 = \sum_{i=1}^{K} \left\{ W(\varpi_i)[H(\varpi_i) - H_d(\varpi_i)] \right\}^2, \tag{8.6}$$

where $H(\varpi_i)$ is the realized FIR's complex frequency response, $H_d(\varpi_i)$ is the desired complex frequency response, and $W(\varpi_i) \geq 0$ is the error weight at the ith frequency locations $\varpi = \varpi_i$. The error weight $W(\varpi_i)$ is a feature that differentiates the LMS method from the window method. The weights can be set high over critically important frequency regions and low in those cases where the response is more or less arbitrary (i.e., don't care). In addition, it has been previously established that an Nth-order linear phase FIR has either even or odd coefficient symmetry. For $N = N_0 + 1$, the general form of an FIR transfer function having an impulse response beginning at sample index $n = 0$, takes the form $H(z) = \sum h[n]z^{-n}$, $n \in [0, N_0]$. There are four types of linear phase cases to be considered and they are as follows:

Type	Order N	Coefficient Symmetry
1	Odd	Even
2	Even	Even
3	Odd	Odd
4	Even	Odd

The frequency response of each type is given by $H(e^{j\varpi}) = Q(\varpi)H^R(\varpi)$, where $H^R(\varpi)$ is a real function of ϖ. The individual filter types satisfy the following relationships:
Type 1:

$$H\left(e^{j\varpi}\right) = e^{-jN_0\varpi/2} H^R(\varpi);$$

$$H^R(\varpi) = h[N_0/2] + 2\sum_{n=1}^{N_0/2} h[(N_0/2) - n]\cos(\varpi n) \tag{8.7}$$

Type 2:

$$H\left(e^{j\varpi}\right) = e^{-jN_0\varpi/2}H^R(\varpi);$$

$$H^R(\varpi) = 2\sum_{n=1}^{(N_0+1)/2} h\left[((N_0+1)/2)-n\right]\cos(\varpi(n-1/2)) \tag{8.8}$$

Type 3:

$$H\left(e^{j\varpi}\right) = je^{-jN_0\varpi/2}H^R(\varpi);$$

$$H^R(\varpi) = 2\sum_{n=1}^{N_0/2} h\left[(N_0/2)-n\right]\sin(\varpi n) \tag{8.9}$$

Type 4:

$$H\left(e^{j\varpi}\right) = je^{-jN_0\varpi/2}H^R(\varpi);$$

$$H^R(\varpi) = 2\sum_{n=1}^{(N_0+1)/2} h\left[((N_0+1)/2)-n\right]\sin(\varpi(n-1/2)) \tag{8.10}$$

To illustrate, the (real) amplitude frequency response of a Type 1 linear phase FIR of order $N = (N_0 + 1)$ (odd) is modeled as

$$|H(\varpi)| = \sum_{k=0}^{M} a_k \cos(\varpi k); \; a_0 = h[M], \, a_i = 2h[i], \tag{8.11}$$

where $M = N_0/2$ (an integer). The parameter M locates the filter's center tap. In a matrix–vector framework, the LMS problem can be represented as

$$H = \begin{bmatrix} W(\varpi_1) & W(\varpi_1)\cos(\varpi_1) & \cdots & W(\varpi_1)\cos(M\varpi_1) \\ W(\varpi_2) & W(\varpi_2)\cos(\varpi_2) & \cdots & W(\varpi_2)\cos(M\varpi_2) \\ \cdots & \cdots & \cdots & \cdots \\ W(\varpi_K) & W(\varpi_K)\cos(\varpi_K) & \cdots & W(\varpi_K)\cos(M\varpi_K) \end{bmatrix};$$

$$a = \begin{bmatrix} a_0 \\ a_1 \\ \cdots \\ a_M \end{bmatrix}; \; d = \begin{bmatrix} W(\varpi_1)H_d(\varpi_1) \\ W(\varpi_2)H_d(\varpi_2) \\ \cdots \\ W(\varpi_K)H_d(\varpi_K) \end{bmatrix}. \tag{8.12}$$

Defining the approximation error e to be $e = Ha - d$, the design objective is to determine the vector that minimizes the error criteria

$$\sigma = e^T e = |e|^2 = (Ha-d)^T(Ha-d) \geq 0. \tag{8.13}$$

The optimizing vector a can be obtained by setting $\partial\sigma/\partial\mathbf{a} = 0$, which results in the minimal squared error solution that is given in terms of the normal equation (also known by other names):

$$H^T Ha = H^T d \Rightarrow a = (H^T H)^{-1} H^T d. \qquad (8.14)$$

Example: LMS Design

Design an LMS fourth-order nonlinear phase FIR having a desired magnitude frequency response $|H(e^{j0})| = 1$, $|H(e^{j2\pi/6})| = 1$, $|H(e^{j2\pi/3})| = 0$, and $|H(e^{j2\pi/2})| = 0$. Choosing uniform error weights, $W(e^{j0}) = W(e^{j2\pi/6}) = W(e^{j2\pi/3}) = W(e^{j2\pi/2}) = 1.0$, Equation 8.12 becomes

$$\mathbf{Hh} = \begin{vmatrix} 1 & 1 & 1 & 1 \\ 1 & \cos(2\pi/6) & \cos(4\pi/6) & \cos(6\pi/6) \\ 1 & \cos(2\pi/3) & \cos(4\pi/3) & \cos(6\pi/3) \\ 1 & \cos(2\pi/2) & \cos(4\pi/2) & \cos(6\pi/2) \end{vmatrix} \begin{bmatrix} h[0] \\ h[1] \\ h[2] \\ h[3] \end{bmatrix}$$

$$= \begin{vmatrix} 1 & 1 & 1 & 1 \\ 1 & 0.5 & -0.5 & -1 \\ 1 & -0.5 & -0.5 & 1 \\ 1 & -1 & 1 & -1 \end{vmatrix} \begin{bmatrix} h[0] \\ h[1] \\ h[2] \\ h[3] \end{bmatrix} = \begin{vmatrix} 1 \\ 1 \\ 0 \\ 0 \end{vmatrix} = \mathbf{d}.$$

From Equation 8.14, $h = (H^T H)^{-1} H^T d = [1/2, 2/3, 0, -1/6]^T$. Observe that the resulting fourth-order FIR has a nonlinear phase response due to a lack of impulse response symmetry. For linear phase designs, impulse response symmetry is required. The magnitude frequency response of the designed FIR is shown in Figure 8.2 for a uniformly error-weighted response. While having the general shape of the desired FIR, the errors are proportionally higher around $\varpi = 2\pi/6$ and $\varpi = 2\pi/3$ (transition band edges). These errors can be reduced by applying a large error weight at these frequencies. Suppose $W(e^{j0}) = W(e^{j2\pi/2}) = 1.0$, and $W(e^{j2\pi/6}) = W(e^{j2\pi/3}) = 2$, a strategy that considers the errors at $\varpi = \pi/3$ and $2\pi/3$ to be twice as egregious as those errors at $\varpi = 0$ and π. The algebraic framework for the weighted design problem now becomes

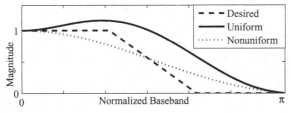

Figure 8.2 LMS FIR designs for weighted (nonuniform) and unweighted (uniform) filters.

$$
\mathbf{Hh} = \begin{vmatrix} W(1) & W(1) & W(1) & W(1) \\ W(2) & \cos(2\pi/6)W(2) & \cos(4\pi/6)W(2) & \cos(6\pi/6)W(2) \\ W(3) & \cos(2\pi/3)W(3) & \cos(4\pi/3)W(3) & \cos(6\pi/3)W(3) \\ W(4) & \cos(2\pi/2)W(4) & \cos(4\pi/2)W(4) & \cos(6\pi/2)W(4) \end{vmatrix} \begin{bmatrix} h[0] \\ h[1] \\ h[2] \\ h[3] \end{bmatrix}
$$

$$
= \begin{vmatrix} 1 & 1 & 1 & 1 \\ 2 & 1 & -1 & -2 \\ 2 & -1 & -1 & 2 \\ 1 & -1 & 1 & -1 \end{vmatrix} \begin{bmatrix} h[0] \\ h[1] \\ h[2] \\ h[3] \end{bmatrix} = \begin{vmatrix} 1 \\ 1 \\ 0 \\ 0 \end{vmatrix} = \mathbf{d}.
$$

Solving, $h = (H^{T}H)^{-1}H^{T}\mathbf{d} = [1/3,\ 1/2,\ 0,\ -1/6]^{T}$. The frequency response shown in Figure 8.2 as the nonuniform weighted solution exhibits a closer fit to the desired response, especially in the transition band. It can also be noted that both FIR have a nonlinear phase response due to an asymmetric impulse response.

MATLAB LMS DESIGN

The MATLAB function firls designs a linear phase FIR filter that minimized the weighted accumulated square error between a desired piecewise linear magnitude frequency response and the realized magnitude frequency response over a set of specified frequencies. The MATLAB firls function

```
h=firls(n,f,a)
```

returns a row vector h containing the $n + 1$ coefficients of an FIR filter whose magnitude frequency response approximates the desired response over frequency range f with gain profile a. The filter's coefficients satisfy the symmetry conditions

$$h[k] = h[n-k], k = 0, 1, \dots, n$$

for a Type 1 FIR, if n is even, and Type 2 FIR, if n is odd. Again it should be noted the $(n + 1)$-order FIR is classified as an nth-order FIR by MATLAB.

Example: LMS Design

The firls function can be used to synthesize a fifth-order (order 4 in MATLAB) Type 1 low-pass linear phase FIR whose magnitude frequency response is that of an ideal piecewise constant filter. The target unit gain passband ranges over $f \in [0, 0.125f_s]$. The zero gain stopband ranges over $f \in [0.\,25f_s, 0.5f_s]$. In MATLAB, these requirements correspond to normalized frequencies [0, 0.25, 0.5, 1] where 1 refers to the Nyquist frequency. To suppress the filter response in the middle of the long stopband $f \in [0.\,25f_s, 0.5f_s]$, a zero gain condition is placed at the midpoint $f = 0.375f_s$. The frequency response of the synthesized FIR is shown in Figure 8.3, and the

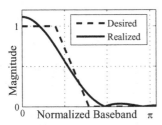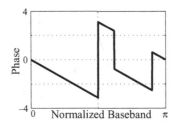

Figure 8.3   Magnitude frequency response (left) and linear phase response in radians (right).

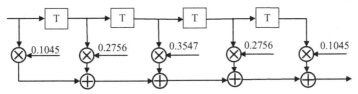

Figure 8.4   Implementation of fifth-order firls linear phase FIR.

physical implementation is shown in Figure 8.4. The quality of the response of the synthesized filter is directly proportional to the filter's order.

Constrained LMS FIR filters can be designed using MATLAB's fircls and fircls1 functions. The fircls syntax is

$$h = \text{fircls}(n-1, f, a, DEV\_UP, DEV\_LO),$$

where the response of the synthesized nth-order filter (MATLAB order $n-1$) approximates the desired filter's piecewise constant frequency response in an LMS sense. The desired filter response is parameterized in terms of f (an array of frequencies) and a (an array of gains). Gain constraints can be imposed above and below the ideal responses and are defined in terms of the parameters DEV_UP and DEV_LO.

Example: Constrained LMS Design

To illustrate a constrained LMS FIR design, consider a 21st-order (order 20 in MATLAB) linear phase low-pass FIR having a normalized passband extending out to $f_p = 0.15f_s$. Note that the stopband edges are not specified. The study begins with a loosely constrained filter and terminates with a more tightly constrained filter. Specifically:

FIR	Passband Gain	Allowable Deviation	Stopband Gain	Allowable Deviation
LMS 1	1	[−0.5, 0.5]	0	[−0.5, 0.5]
LMS 2	1	[−0.5, 0.025]	0	[0.5, 0.025]
LMS 3	1	[−0.025, 0.025]	0	[−0.025, 0.025]

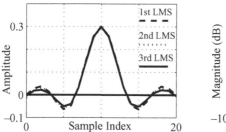 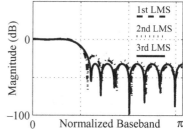

Figure 8.5   Low-pass constrained linear phase fircls filter. Note the similarity of the impulse responses (left) and the dissimilar magnitude frequency responses due to the imposed constraints (right).

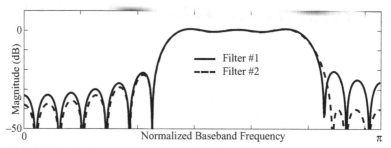

Figure 8.6   Magnitude frequency response of band-pass FIR implemented using fircls.

The results are presented in Figure 8.5. The derived impulse responses appear to be very similar. The magnitude frequency responses (in decibel) do, however, exhibit some points of differentiation. LMS 1 is essentially an unconstrained filter since all the passband and stopband error deviations fall well within the ±0.5 requirement. LMS 2 tightens up the passband and, in doing so, gives back some stopband attenuation resulting in the minimum stopband attenuation that is less than that of LMS 1. The more restrictive LMS 3 has to fight to meet the specifications resulting in, for example, minimum stopband gains that are essentially constant across the stopband in order to meet the designer's requirements. When the design is complete, the stopband edge frequency will need to be separately computed.

Example: fircls LMS Design

Design a 31st-order linear phase band-pass FIR having a normalized passband of f_P ranging from $0.2f_s$ to $0.4f_s$. Note that the stopband edges are not specified. Begin with a uniformly constrained filter followed by an FIR with unsymmetrical constraints. The first FIR has a uniform 10% constraint on the first and second stopbands and passband. The second FIR has a relaxed first stopband, tightened second stopband, and the original 10% passband. The effects of these changes can be seen in the magnitude frequency response shown in Figure 8.6. Notice that even though the first stopband constraints were relaxed, they were effactually increased in the final design.

The MATLAB *FIRCLS1* function is used to create an nth-order ($n - 1$ order in MATLAB) low- and high-pass constrained least-squares FIRs using the following syntax:

$$h = \text{FIRCLS1}(n-1, wO, DEVP, DEVS)$$

with a cutoff frequency wO and maximum band deviations or ripples (in linear units) limited by *DEVP* and *DEVS*. For a high-pass design, use the following:

$$h = \text{FIRCLS1}(n-1, wO, DEVP, DEVS, \text{"high"}) \text{ is a high-pass filter.}$$

A low-pass extension is

$$h = \text{FIRCLS1}(n-1, wO, DEVP, DEVS, wP, wS, K),$$

which weighs the square error in the passband K times greater than that in the stopband. Continuing, wP is the passband edge of the L_2 weight function and wS is the stopband edge ($wP < wO < wS$).

Example: fircls1 LMS Design

Design a 31st-order (30 in MATLAB) linear phase low-pass FIR having a normalized passband of $f_p = 0.15\pi$. The fircls1 routine can be based on the use of L_2 norms and can set "soft" passband and stopband edges. The design study begins with a loosely constrained filter, followed by a more highly constrained filter. Note how the constraints also affect the transition bandwidth as illustrated in Figure 8.7.

MATLAB DESIGN COMPARISONS

MATLAB provides several tools that can be used to create LMS and other classes of FIR filters. The fir2 window design method is used to design a linear phase frequency sampling FIR using a selected window (default Hamming). The fircls function produces a constrained (gain) LMS design. The filter routine fircls1 produces a weighted constrained (gain) low-pass or high-pass LMS design. Program firls also produces a least-squares FIR, which can provide baseline services as well as operate as a differentiator or Hilbert filter. To compare these design strategies, a

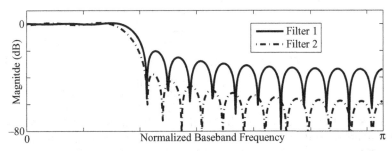

Figure 8.7   Magnitude frequency response of low-pass FIR implemented using fircls1.

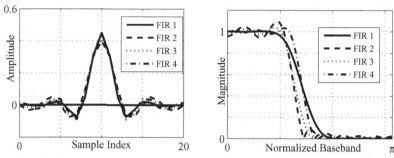

Figure 8.8   Comparison of linear phase LMS FIR design strategies. Impulse response (left) and magnitude frequency response (right).

family of 21st-order FIR models (order 20 in MATLAB) is constructed. These 21st-order FIRs, using fir2, fircls, and fircls1, are generated, and the FIRs' impulse responses are shown in Figure 8.8 along with their magnitude frequency responses. Referring to Figure 8.8, it can be observed that the first FIR filter is a 21st-order fir2 low-pass filter with a normalized passband cutoff frequency of 0.2π and stopband cutoff of 0.25π, using a default Hamming widow. The second FIR is a 21st-order fircls low-pass FIR with a passband cutoff frequency of 0.2π, 0.1 passband deviation, and 0.1 stopband deviation. The third FIR is a 21st-order fircls1 low-pass FIR with cutoff frequency of 0.2π, 0.2 passband deviation out to 0.175π, and 0.2 stopband deviation beginning at 0.25π, with a weighting factor of $k = 10$. The fourth FIR is a 21st-order firls low-pass FIR with a passband cutoff frequency of 0.2π and stopband cutoff of 0.25π. It can be noted that there is, at a macrolevel, general agreement among the filter trials. However, they do differ at the microlevel. Some may meet a given set of specifications, some may not.

PRONY'S METHOD

A question logically arises regarding how to design an FIR from only an empirical or experimental understanding of a filter's response. In such cases, Prony's method can sometimes be used to map a measured time series $h[k] = \{h_0, h_1, h_2, \ldots\}$ into an equivalent transfer function having the form

$$H(z) = \sum_{k=-\infty}^{\infty} h_k z^{-k} = \frac{\displaystyle\sum_{k=0}^{N} b[k]z^{-k}}{1 + \displaystyle\sum_{k=0}^{M} a[k]z^{-k}}. \tag{8.15}$$

Equation 8.15 does not automatically specify an FIR due to the terms appearing in the denominator. If the denominator terms are removed, then the transfer function has an FIR form

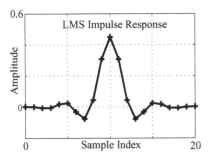
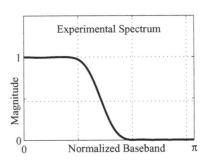

Figure 8.9   Prony's database. First 21 samples (left) and resulting magnitude frequency response (right).

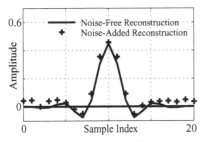
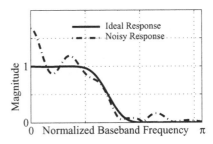

Figure 8.10   Application of Prony's method. First 21 noise-contaminated samples (left) and resulting magnitude frequency response (right).

$$H(z) = \sum_{k=0}^{N} b[k]z^{-k} \tag{8.16}$$

Such a filter would have an impulse response of length $N + 1$. It should be apparent that the first $N + 1$ samples immediately define the $N + 1$ coefficients $b[k]$, $k \in [0, N]$. Prony's method automatically performs this task, even in the presence of noise. In such cases, Prony's method will produce the best LMS estimate of the monitored FIR reponse.

Example: Prony's Method

The impulse response of a 21st-order low-pass linear phase FIR is observed over 31 consecutive sample instances. The observed FIR's impulse response and derived magnitude frequency response are shown in Figure 8.9. The impulse response could have been immediately deduced from the measured impulse response without the intervention of Prony's method. Suppose, however, that the measured impulse response samples are contaminated by additive noise. In such instances, the data shown in Figure 8.10 will result. The additive noise is seen to have an effect on the resulting spectrum.

CHAPTER 9

EQUIRIPPLE DESIGN METHOD

EQUIRIPPLE CRITERION

The least mean square (LMS) filter synthesis methodology can produce a design having potentially large localized errors. An alternative design strategy is one that will insure that the largest approximation error is bounded below some acceptable value. This is the role of the equiripple design rule that satisfies what is called a minimax error criteria given by

$$\delta(\varpi_e) = \text{minimize}\{\text{maximum}(\varepsilon(\varpi) \mid \varpi \in [0, \varpi_s/2)\}, \tag{9.1}$$

where $\delta(\varpi_e)$ is called the minimax or extremal error where the error at frequency ϖ is defined to be

$$\varepsilon(\varpi) = W(\varpi)\left|H_d\left(e^{j\varpi}\right) - H\left(e^{j\varpi}\right)\right| \tag{9.2}$$

with $W(\varpi) \geq 0$, a non-negative error weight. It can be noted that the error $\varepsilon(\varpi)$ is the weighted difference between the desired $H_d(e^{j\varpi})$ and realized filter's response $H(e^{j\varpi})$ at the normalized baseband frequency ϖ. An Nth-order minimax FIR design strategy minimizes the maximum absolute error over all baseband of frequencies. These maximum errors $\delta(\varpi_e)$, called extremal errors, occur at the extremal frequencies ϖ_e. The extremal errors satisfy $|\delta(\varpi_e)| \geq |\delta(\varpi)|$ across the baseband. In addition, if $\delta(\varpi_i)$ and $\delta(\varpi_k)$ are extremal errors measured at the extremal frequencies ϖ_i and ϖ_k, then $|\delta(\varpi_i)| = |\delta(\varpi_k)|$ for all i and k. In addition, the signs of the extremal errors are also known to alternate between adjacent extremal frequencies so that $\delta(\varpi_i) = -\delta(\varpi_{i+1})$. Because all the extremal errors are equal in magnitude and alternating in sign, the resulting filter is also referred to as an equiripple finite impulse response (FIR).

It has been previously established that an Nth-order linear phase FIR has either even or odd coefficient symmetry. For $N = N_0 + 1$, the general form of an Nth FIR having an impulse response, beginning at sample index $n = 0$, has a transfer function given by

Digital Filters: Principles and Applications with MATLAB, First Edition. Fred J. Taylor.
© 2012 by the Institute of Electrical and Electronics Engineers, Inc.
Published 2012 by John Wiley & Sons, Inc.

$$H(z) = \sum_{n=0}^{N_0} h[n] z^{-n} \tag{9.3}$$

The four types of linear phase FIRs are the following:

Type	Order N	Coefficient Symmetry
1	Odd	Even
2	Even	Even
3	Odd	Odd
4	Even	Odd

The frequency response of each type is given by $H(e^{j\omega}) = (j(\omega))H^{R}(\omega)$, where $H^{R}(\omega)$ is a real function of ω. In this context, the individual filter types satisfy

Type 1:
$$H(e^{j\omega}) = e^{-jN_0\omega/2} H^R(\omega);$$
$$H^R(\omega) = h[N_0/2] + 2\sum_{n=1}^{N_0/2} h[(N_0/2) - n]\cos(\omega n) \tag{9.4}$$

Type 2:
$$H(e^{j\omega}) = e^{-jN_0\omega/2} H^R(\omega);$$
$$H^R(\omega) = 2\sum_{n=1}^{(N_0+1)/2} h[((N_0+1)/2) - n]\cos(\omega(n-1/2)) \tag{9.5}$$

Type 3:
$$H(e^{j\omega}) = je^{-jN_0\omega/2} H^R(\omega);$$
$$H^R(\omega) = 2\sum_{n=1}^{N_0/2} h[(N_0/2) - n]\sin(\omega n) \tag{9.6}$$

Type 4:
$$H(e^{j\omega}) = je^{-jN_0\omega/2} H^R(\omega);$$
$$H^R(\omega) = 2\sum_{n=1}^{(N_0+1)/2} h[((N_0+1)/2) - n]\sin(\omega(n-1/2)) \tag{9.7}$$

where the real valued function $H^R(\omega)$ exhibits the sign changes that characterize an equiripple solution. The standard means of achieving an equiripple solution is the Remez exchange algorithm.

REMEZ EXCHANGE ALGORITHM

For a linear phase FIR, the location of the maximum errors can be numerically computed using the alternation theorem from polynomial approximation theory. This method, popularized by Parks and McClellan, is used to iteratively solve a system of equations using what is called the Remez exchange algorithm. The Remez exchange algorithm continually adjusts the tentative location of an extremal frequency ϖ_i until the minimax criterion is satisfied in terms of an ε stopping rule. This method has been used for decades to design linear phase FIRs and continues to provide reliable results.

The standard equiripple design method, based on the Remez exchange algorithm, can be motivated in terms of the design of a physically realizable Nth-order Type 1 linear phase FIR. It is assumed that the filter's amplitude frequency response (i.e., $H^R(\varpi)$) can be compactly modeled as a real function of ϖ, denoted $A(\varpi)$, and has the form

$$A(\omega) = \sum_{i=0}^{L} a_i \left(\cos \left(i\varpi \right) \right); a_0 = h[L]; a_i = 2h[i] \tag{9.8}$$

where $L = N_0/2$ (an integer) locates the FIR's center tap. The desired FIR magnitude frequency response, measured at frequency ϖ_i, is assumed to be $D(\varpi_i) = |H_d(\varpi_i)|$ where $H_d(\varpi_i)$ is the desired (positive baseband) frequency response measured at ϖ_i. The equiripple design method determines the $L + 1$ filter coefficients that satisfy the minimax error criterion. The error process is modeled using Chebyshev polynomials that insure the extremal errors alternate in sign (i.e., $\pm\delta$). The optimizing filter can be modeled in a linear algebraic sense in terms of $L + 2$ equations in $L + 2$ unknowns. Specifically, for $a_i = a[i]$,

$$Ha = \begin{bmatrix} 1 & \cos(\varpi_0) & \dots & \cos(L\varpi_0) & 1 \\ 1 & \cos(\varpi_1) & \dots & \cos(L\varpi_1) & -1 \\ \dots & \dots & \dots & \dots & \dots \\ 1 & \cos(\varpi_L) & \dots & \cos(L\varpi_L) & (-1)^L \\ 1 & \cos(\varpi_{L+1}) & \dots & \cos(L\varpi_{L+1}) & (-1)^{L+1} \end{bmatrix} \begin{bmatrix} a[0] \\ a[1] \\ \dots \\ a[L] \\ \delta \end{bmatrix} = \begin{bmatrix} D(\varpi_0) \\ D(\varpi_1) \\ \dots \\ D(\varpi_L) \\ D(\varpi_{L+1}) \end{bmatrix} = d, \tag{9.9}$$

where $(-1)^r$ is the result of the fact that the extremal errors alternate in sign. The equiripple method iterates until a minimax error criterion is satisfied, a condition where $A(\varpi_i) - D(\varpi_i) = (-1)^i \delta$ evaluated at the extremal frequencies ϖ_i, $i \in [0, L + 1]$ within some ε stopping criterion. Once the coefficient set $a[i] = 2h[i]$, $a[0] = 2h[0]$, has been determined, the Type 1 filter coefficient space is filled using the symmetry assignment rule $h[i] = h[-i]$.

A variation on this theme assigns individual weights to the critical frequencies to achieve a weighted minimax error, producing $(A(\varpi_i) - D(\varpi_i))W(\varpi_i) = (-1)^i \delta$ at the extremal frequencies ϖ_i, $i \in [0, L + 1]$, and $W(\varpi_i) \geq 0$. The weighted minimax problem can be casted into the linear algebraic framework shown below:

$$
H_W h = \begin{bmatrix}
1 & \cos(\varpi_0) & \cdots & \cos(L\varpi_0) & 1/W(\varpi_0) \\
1 & \cos(\varpi_1) & \cdots & \cos(L\varpi_1) & -1/W(\varpi_1) \\
\cdots & \cdots & \cdots & \cdots & \cdots \\
1 & \cos(\varpi_L) & \cdots & \cos(L\varpi_L) & (-1)^L/W(\varpi_L) \\
1 & \cos(\varpi_{L+1}) & \cdots & \cos(L\varpi_{L+1}) & (-1)^{L+1}/W(\varpi_{L+1})
\end{bmatrix}
\begin{bmatrix}
a[0] \\
a[1] \\
\cdots \\
a[L] \\
\delta
\end{bmatrix}
$$

$$
= \begin{bmatrix}
D(\varpi_0) \\
D(\varpi_1) \\
\cdots \\
D(\varpi_L) \\
D(\varpi_{L+1})
\end{bmatrix} = d.
\tag{9.10}
$$

The Remez exchange algorithm produces a minimax solution, which is as close to being optimal up to the limits of an ε stopping criteria. The computational process is abstracted below for an Nth-order FIR design, where $N = N_0 + 1$ and $L = N_0/2$:

1. Initially set the $(L + 2)$ candidate extremal frequencies ϖ_i to arbitrary values.
2. Determine the approximation error value:

$$
\delta = \frac{c_0 D(\varpi_0) + c_1 D(\varpi_1) + \ldots + c_{L+1} D(\varpi_{L+1})}{\dfrac{c_0}{W(\varpi_0)} - \dfrac{c_1}{W(\varpi_1)} + \ldots (-1)^{L+1} \dfrac{c_{L+1}}{W(\varpi_{L+1})}},
\tag{9.11}
$$

where

$$
c_n = \prod_{\substack{i=0 \\ i \neq n}}^{L+1} \frac{1}{\cos(\varpi_n) - \cos(\varpi_i)}.
\tag{9.12}
$$

3. The values of the realized magnitude response $A(\varpi)$ at $\varpi = \varpi_i$ are computed as

$$
A(\varpi_i) = \frac{(-1)^i \delta}{W(\varpi_i)} + D(\varpi_i), \quad 0 \leq i \leq L+1.
\tag{9.13}
$$

4. The polynomial $A(\varpi)$ is determined by interpolating the above equation at the $L + 2$ extremal frequencies using the Lagrange interpolation formula:

$$
A(\varpi) = \sum_{i=0}^{L+1} A(\varpi_i) P_i(\cos(\varpi)),
\tag{9.14}
$$

where

$$P_i(\cos(\varpi)) = \prod_{\substack{j=0 \\ j \neq i}}^{L+1} \frac{\cos(\varpi) - \cos(\varpi_j)}{\cos(\varpi_i) - \cos(\varpi_j)}, 0 \leq i \leq L+1. \tag{9.15}$$

5. The new weighted error function is computed within a dense set $S(S \geq L)$ of frequencies ($S = 16L$ typical) to determine the $L + 2$ new extremal frequencies from the error criterion values.

6. If the peak error values of δ are nearly equal in absolute value, terminate; otherwise return to Step 2.

Example: Equiripple Criterion

The objective is to synthesize a linear equation $A(\varpi) = a_1\varpi + a_0$ solution that best fits a desired quadratic profile $D(\varpi) = 1.1\varpi^2 - 0.1$, for $\varpi \in [0, 2]$ in a minimax sense. Let the initial guess of the extremal frequency locations be $\varpi = 0$, 0.5, and 1.5, corresponding to desired filter gains of $D(\varpi) = -0.1$, 0.175, 0.175, and 2.375, respectively. For uniform error weighting, $A(\varpi)$ has a minimax fit to $D(\varpi)$ that is obtained using the computer pseudocode below:

```
f=rmp(101,2/100,101)  # ramp frequency [0, 2]

d=1.1*f^2-0.1  # desired response

A={[1,0,1],[1,0.5,-1], [1,1.5,1]}  # A(ϖ)=a1ϖ+a0+□

v={-0.1,0.175,2.375}  # desired response

a=inv(A)*v  # solution

a

    -0.375  # a0

     1.65   # a1

     0.275  # ε

Aw=1.65*f-0.375  # Current value of A(ϖ)=a1ϖ+a0
```

Finally, $D(\varpi)$, $A(\varpi)$, and $\varepsilon(\varpi)$ are shown in Figure 9.1. Note that the errors at the extremal test frequencies are the same ($\varepsilon = 0.275$), as required of a minimax solution. However, the tested frequencies are not the actual minimizing extremal frequencies. The search continues.

Choose new frequencies that are found at the extreme error locations of Figure 9.1. They are $\varpi = 0$, 0.75, and 2 and correspond to desired filter gains $D(\varpi)$ of -0.1, 1.0, 0.5188, and 4.3, respectively. For uniform error weighting, the next iteration produces the following:

```
A={[1,0,1],[1,0.75,-1],[1,2,1]}  # A(ϖ)=a1ϖ+a0+ε

v={-0.1,0.5188,4.3}   # desired response

a = inv(A)*v  # solution
```

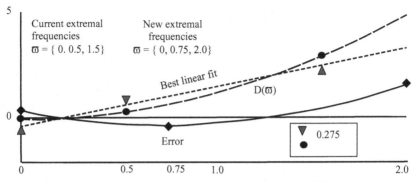

Figure 9.1 First iteration outcome with the maximal error at $\varpi = 0.75$.

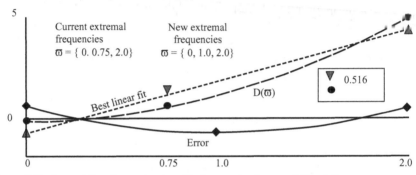

Figure 9.2 Second iteration outcome with the maximal error at $\varpi = 1.0$.

```
a

   -0.616  # a0

    2.2  # a1

   0.516  # ε

Aw=2.2*f-0.6156  # Current value of A(ϖ)=a1ϖ+a0
```

Finally, $D(\varpi)$, $A(\varpi)$, and $\delta(\varpi)$ are shown in Figure 9.2. Note that the errors at the test frequencies are the same ($\varepsilon = 0.516$) and differ from the first trial ($\varepsilon = 0.275$). Again, the selected frequencies are not the actual extremal frequencies.

Choose the new candidate frequencies found in Figure 9.2. They are $\varpi = 0$, 1, and 2, respectively with desired filter gain $D(\varpi)$ of -0.1, 1.0, and 4.3, respectively. For uniform error weighting, the next iteration produces the following:

```
A={[1,0,1],[1,1,-1],[1,2,1]}  # A(ϖ)=a1ϖ+a0+ε

v={-0.1,1.0,4.3}     # desired response

a=inv(A)*v  # solution

a

   -0.65  # a0
```

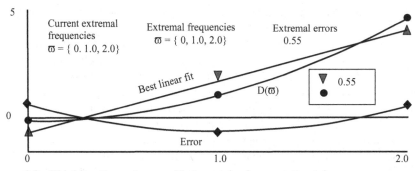

Figure 9.3 Third iteration outcome with the maximal error at $\varpi = 1.0$.

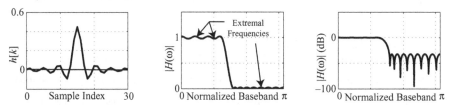

Figure 9.4 Type 1 31st-order FIR equiripple approximation to a piecewise linear FIR magnitude frequency response model showing (left) the impulse response, (middle) the magnitude frequency response, and (right) the magnitude frequency response in decibels. Also shown are the equiripple extremal errors $|\delta| = |\delta_p| = |\delta_a|$ at the extremal frequency locations.

```
        2.2    # a1

        0.55   # ε

Aw=2.2*f-0.65   # Current value of A(ϖ)=a1ϖ+a0
```

Finally, $D(\varpi)$, $A(\varpi)$, and $\varepsilon(\varpi)$ are shown in Figure 9.3. For the final iteration, the extremal frequencies are unchanged and have a minimax error value of $\varepsilon = 0.55$. Therefore, upon completion of the third iteration, a solution is found with unchanged extremal frequencies.

A typical equiripple FIR approximation of a piecewise constant ideal filter model is presented in Figure 9.4. The equiripple design objectives can be seen to be satisfied since the extremal passband and attenuation (stop) band deviations (errors) are equal. The reason that no extremal frequencies are located in the transition band is that no transition band performance requirements were specified.

WEIGHTED EQUIRIPPLE FIR DESIGN

The uniformly weighted ($W(\varpi) = 1$) solution that exhibits equal extremal errors at extremal frequencies. There are times, however, when it is desired to independently specify the pass and stopband ripple deviations parameters δ_p and δ_a. This is the role of the non-negative error weight $W(\varpi) \geq 0$. If $W_{\text{passband}}(\varpi) = W_{\text{stopband}}(\varpi)$, then the error deviations found in the passband are treated the same as those in the stopband. The realized filter will produce a magnitude frequency response that satisfies $\delta = \delta_p = \delta_a$. Such solutions are said to be uniformly weighted. Choosing a weighted relationship $W_p(\varpi)$ (passband) $= 10W_a(\varpi)$ (attenuation band) implies that the extremal error in the passband is to be considered to be 10 times more serious (undesirable) than those found in the attenuation band. The resulting solution to the weighted equiripple error minimization problem would be a solution in which $10\delta_p = \delta_a$. That is, the actual filter error deviation in the passband would be 1/10th that of the stop or attenuation band. In this manner, significant to subtle adjustments can be made to the magnitude frequency response envelope of an equiripple design. Finally, the rms value of the error deviation is given by $\delta = \sqrt{\delta_p \delta_a}$ where δ is the error for a uniformly weighted design (i.e., $W(\varpi) = 1$). To illustrate, consider a 51st-order low-pass equiripple FIR having uniform error weights to meet the following specifications:

- sampling frequency $f_s = 100$ khz;
- frequency band 1: $f \in [0.0, 10]$ khz, desired gain $= 1.0$, $W(f) = 1$ (passband);
- frequency band 2: $f \in [15, 50]$ khz, desired gain $= 0.0$, $W(f) = 1$ (stopband).

The resulting equiripple filter is displayed in Figure 9.5. The resulting Type 1 filter has a worst case gain computed to be $G_{max} = 1.67565 < 2^1$. The measured minimax error is $\delta = \delta_p = \delta_a = 0.00428$ or -47.4 dB, and $\delta = \sqrt{\delta_p \delta_a}$. By decreasing the filter order, the deviation error will increase, and increasing the order will decrease the error deviation. It should be appreciated, however, that filter order is directly correlated to the number of multiply-accumulate (MAC) cycles required per filter cycle. This is inversely related to the maximum run-time bandwidth of the filter. This constraint can often be used to establish the maximum accepted filter order.

The 51st-order Type 1 low-pass equiripple FIR that has been designed using uniform error weights has a -47.4-dB passband and stopband error deviation. The -47.4-dB passband deviation (± 0.00428) would, in most instances, be considered to be excessive for most applications. A passband deviation of ± 0.1 or -20 dB is considered to be more realistic. This redefinition of the passband maximal attenuation can be achieved using sub-band weights in an intelligent manner. Assigning the stopband weight to be 1000 times that of the passband weight (i.e., 1000

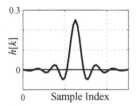 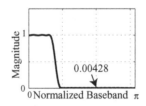 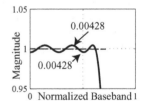

Figure 9.5   Fifty-first-order low-pass FIR using uniform passband and stopband weights. Impulse response (left), magnitude frequency response (middle), and zoom expanded magnitude frequency response (right) along with the direct measurement of the passband extremal errors.

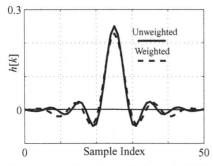
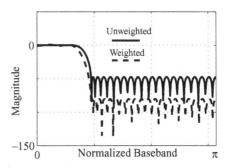

Figure 9.6   Fifty-first-order low-pass FIR using uniform and nonuniform passband and stopband weights (1:1000). Impulse responses (left); magnitude frequency responses in decibels (right).

$W_{passband}(\omega) = W_{stopband}(\omega))$ results in the FIR shown in Figure 9.6. The passband deviation is measured to be approximately $\delta_p = \pm 0.0954302 \sim (\pm 0.1)$ and the stopband weight is estimated to be $\delta_a = -79$ dB. This is a relaxation factor greater than 20 in the passband and corresponds to an increase of 27.5 dB. That is, $\delta_p \sim 0.1$, $\delta_a \sim 0.00018$, and $\delta = \sqrt{\delta_p \delta_a} = 0.00424264$, which is essentially the minimax error for the uniformly weighted value of δ. Finally, the worst case gain of the filter is computed to be $G_{max} = 1.72627 < 2^1$, which is a slight increase over the uniformly weighted case.

Example: Equiripple FIR

A 51st-order band-pass equiripple Type 1 FIR is designed to meet the following specifications:

- sampling frequency $f_s = 100$ kHz;
- frequency band 1: $f \in [0.0, 10]$ kHz, desired gain = 0.0, $W(f) = 1$ (stopband #1);
- frequency band 2: $f \in [12, 38]$ kHz, desired gain = 1.0, $W(f) = 1$ (passband);
- frequency band 3: $f \in [40, 50]$ kHz; desired gain = 0.0, $W(f) = 1$ (stopband #2).

The resulting filter is shown in Figure 9.7.

The impulse response is seen to have even symmetry and the magnitude frequency response exhibits equal equiripple deviation from the ideal in both the passband and stopband. The measured extremal errors are $\delta_p = \delta_a = 0.0664$ (−23.5 dB). The 0.0664 passband deviation may be considered to be too restrictive (overspecified). A $\delta_p = 0.125$ passband deviation is considered to be more realistic. This can be achieved using sub-band weights in an intelligent manner. The required weights can be determined experimentally for an FIR having the following design specifications:

- sampling frequency $f_s = 100$ kHz;
- frequency band 1: $f \in [0.0, 10]$ kHz, desired gain = 0.0, $W(f) = 4$;

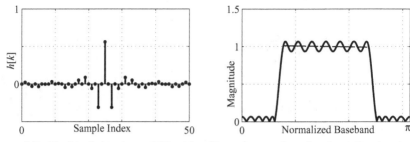

Figure 9.7 Equiripple band-pass FIR with uniform pass- and stopband weights showing impulse response (left) and magnitude frequency response (right).

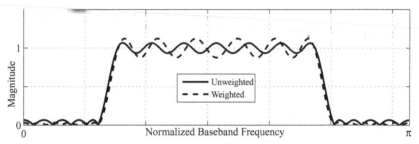

Figure 9.8 Fifty-first-order equiripple band-pass FIRs designed using uniform and nonuniform weights. Observe the trade-off of passband gain against stopband attenuation for the nonuniformly weighted case.

- frequency band 2: $f \in [12, 38]$ kHz, desired gain = 1.0, $W(f) = 1$;
- frequency band 3: $f \in [40, 50]$ kHz, desired gain = 0.0, $W(f) = 4$.

The FIR is designed using the Remez method. The measured passband and stopband deviations from the ideal are found to be given by $\delta_p = 0.125$ and $\delta_a = 0.033$ (~−30 dB). While the passband deviation has been relaxed to a more acceptable value, the stopband attenuation has increased from −23.5 dB to about −30 dB. This is normally considered to be an acceptable if not a desirable trade-off. For comparative purposes, the weighted and uniform weighted magnitude frequency responses are shown in Figure 9.8. Notice also that $\delta = \sqrt{\delta_p \delta_a} = 0.065$, which is essentially the error deviation of the uniformly weighted design.

HILBERT EQUIRIPPLE FIR

Equiripple filters are commonly used in the design of low-pass, band-pass, band-stop, and high-pass linear phase filters. Another important use of the equiripple design paradigm is to construct FIR filters that can implement Hilbert transforms. Hilbert transforms, or filters, are important elements in many communication systems. A Hilbert filter has a frequency response given by

$$H(e^{j\varpi}) = \begin{cases} j: & -\pi/2 \leq \varpi < 0 \\ -j: & 0 < \varpi \leq \pi/2 \end{cases}. \tag{9.16}$$

Observe that a Hilbert filter is essentially an all-pass filter in that $|H(e^{j\varpi})| = 1$, for all $\varpi \in [-\pi/2, \pi/2)$, but possesses a distinctive quadrature phase-shifting property. The phase-shifting property of a Hilbert filter is particularly useful in defining single sideband modulators, quadrature amplitude modulation (QAM) communications systems, and systems having in-phase and quadrature phase (I/Q) channels. The quadrature phase-shifting ability of a Hilbert filter can be viewed in framework of a pure cosine and sine wave, where $y[k] = \cos(\phi k) = (e^{-j\phi k} + e^{-j\phi k})/2$ and $y[k] = \sin(\phi k) = (e^{-j\phi k} + e^{-j\phi k})/2j$. Assuming that the tone $y[k]$ is in the passband of the Hilbert filter, Equation 9.17 states that

$$\cos(\omega k) = \frac{\delta[\omega - k] + \delta[\omega + k]}{2} \xrightarrow{\text{Hilbert}} \frac{j\delta[\omega - k] - j\delta[\omega + k]}{2} = -\sin(\omega k)$$

$$\sin(\omega k) = \frac{\delta[\omega - k] - \delta[\omega + k]}{2j} \xrightarrow{\text{Hilbert}} \frac{\delta[\omega - k] + \delta[\omega + k]}{2} = \cos(\omega k)$$
$$\tag{9.17}$$

There is a minor problem associated with implementing a practical equiripple Hilbert filter. In particular, implementing the sharp phase transition from $-90°$ to $90°$ at 0 Hz is difficult. This problem can be mitigated by creating a small guard band around 0 Hz (DC) and $f_s/2$.

Example: Hilbert FIR

An ideal Hilbert filter would have a unity baseband gain that cannot be realized by an approximating equiripple FIR. Instead, locally defined guard bands around $f = 0$ and $f = f_s/2$ are asserted. A 63rd-order Hilbert FIR can be designed that meets the following specifications:

- sampling frequency $f_s = 1$ Hz (normalized);
- frequency band 1: $f \in [0.0, 0.01]$ kHz, desired gain = 0.0, $W(f) = 1$ (guard band);
- frequency band 2: $f \in [0.02, 0.48]$ kHz, desired gain = 1.0, $W(f) = 1$ (passband);
- frequency band 3: $f \in [0.49, 0.50]$ kHz, desired gain = 0.0, $W(f) = 1$ (guard band).

The sample frequency was set to unity for convenience. The FIR is designed using the equiripple method with respect to the given gain profile shown in Figure 9.9. The FIR has a passband and stopband deviation of δ_p, which is measured to be approximately $\delta = -15.8039$ dB. The impulse response is also seen to be that of an odd symmetry (Type 3), representing a linear phase FIR.

Differentiator

Equiripple filters are also adept at implementing other filter classes of FIRs, such as a linear phase differentiator of order N. An Nth-order differentiator's frequency

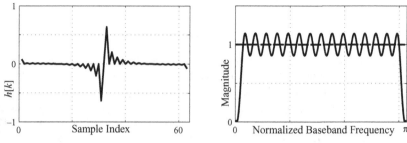

Figure 9.9 Hilbert FIR filter impulse response (left) and magnitude frequency response showing the presence of guard bands (right).

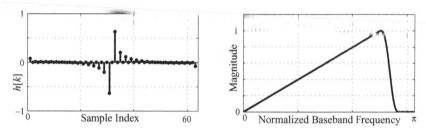

Figure 9.10 Fifty-first-order differentiator's impulse (left) and magnitude frequency response (right).

response is given by $H(e^{j\varpi}) = (j\varpi)^{N}$. For practical realization reasons, the desired frequency response specifications should include a guard band near the Nyquist frequency. That is, it is desired to set $|H(e^{j\varpi})| = 0$ for ϖ near $\varpi = \pi/2$.

Example: Differentiator

A 51st-order linear phase differentiator is to be designed having a differentiating baseband frequency response out to $0.4f_s$ and a stopband buffer beginning at $0.45f_s$. The differentiator's impulse and magnitude frequency responses are shown in Figure 9.10. Notice that the impulse response is antisymmetric (Type 3) and that a stopband guard band has been added, buffering the end of the passband.

EQUIRIPPLE ORDER ESTIMATE

There are several order estimation algorithms commonly used to translate a set of design parameters (δ_p, δ_a, ω_p, ω_a) into an estimate of the equiripple order based on a linear phase model. The most popular estimation formulas are summarized below. The first estimator is given by

$$N_{\text{FIR}} \approx \frac{-10\log_{10}(\delta_a\delta_p)-15}{14\Delta\omega}+1, \tag{9.18}$$

where $\Delta\omega$ is the normalized transition bandwidth and is given by $\Delta\omega = (\omega_a - \omega_p)/\omega_s$. Another order estimator is similar and is given by

$$N_{FIR} \approx \frac{-10\log_{10}(\delta_a\delta_p)-13}{14.6\Delta\omega}+1. \tag{9.19}$$

These formulas provide an estimate of an equiripple filter's order needed to meet linear phase FIR design requirements specified in terms of δ_p, δ_a, ω_p, and ω_a. For a uniform error weighted equiripple FIR (i.e., $W(\varpi) = 1$), $\delta_a = \delta_p = \delta$, and for a nonuniformly weighted equiripple FIR with $\sqrt{\delta_a\delta_p} = \delta$. For example, a six-channel DVD audio player running at 96 kSa/s would require that a 100-tap FIR operate at 60 M MAC/s. The order prediction formula suggests that the design of a steep-skit or narrow-band equiripple FIR will result in an unrealistic high-order solution. As a "rule of thumb," when the normalized transition band $\Delta\varpi$ has a value less than 0.04, an equiripple filter is virtually impossible to build. Another problem with high-order FIRs is that the coefficients found out on the tapers (tap weight coefficients found at the extreme beginning or end of the impulse response) would have such small values that they could not be successfully resolved by a fixed-point word.

Example: FIR Order

Consider again the nonuniformly weighted low-pass equiripple FIR filter studied in an earlier example. In particular, $\delta_p = 0.1$, $\delta_a = 0.00018$, and $\Delta\varpi = 0.05$. From Equation 9.18, one obtains an estimate of $N = 47.3$, and from Equation 9.19, $N = 48.3$, both being consistent with the realized order of $N = 51$.

MATLAB EQUIRIPPLE FIR

MATLAB provides several means of designing equiripple FIRs. The most common of these are as follows:

- firpm (equiripple FIR design using the Parks–McClellan method),
- firpmord (order estimation for equiripple FIR design using the Parks–McClellan method).

Other toolkits, such as the Filter Design Toolbox, include the following:

- firgr (Parks–McClellan equiripple filter),
- fircband (constrained-band equiripple filter),
- firxceqrip (constrained low-pass equiripple filter),
- firequint (equiripple FIR interpolation filter),
- fircgr (Parks–McClellan equiripple filter).

These methods share many core attributes, differing mainly in subtle features. The most versatile tools are the firgr and firpm modules. As with other MATLAB FIR design tools, a traditional nth-order FIR is of order $n - 1$ in MATLAB.

Example: Equiripple MATLAB design

A five-band linear phase Type 1 FIR is defined using MATLAB. Initially, the band error weights are set to unity across the baseband. A second design is requested with the stopband errors being progressively smaller than those of the unity weighted passband. Finally, the process is reversed, strongly weighting the passband errors. The three distinct designs are reported in Figure 9.11. The relative trade-offs between passband and stopband gains as a function of error weights can be clearly seen.

The MATLAB function firpmorder function can be used to estimate the order of a linear equiripple FIR. Use firpmorder to design, for example, a minimum-order low-pass filter with a 500-Hz passband cutoff frequency and 600-Hz stopband cutoff frequency, having a sampling frequency of 2000 Hz, and at least 40-dB attenuation in the stopband with less than 3 dB of ripple in the passband as shown in Figure 9.12. A 23rd-order (order 22 in MATLAB) FIR is required that is implemented and displayed in Figure 9.12.

There is no hard and fast rule that would suggest which class of classic FIR (window, LMS, equiripple) can best meet a set of filter requirements with the lowest order solution. The success of a filter solution is predicated on the decisions made by the filter designer. Nevertheless, one of the most basic decisions to be made is what type of filter best meets the design specifications. There is a common tendency to assume that an equiripple is generally a "best fit." This is not necessarily universally correct. However, with the abundance of filter design software, a family of designs can be rapidly implemented and compared. To illustrate (see Fig. 9.13),

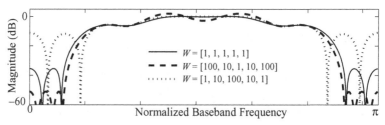

Figure 9.11 Multiple 31st-order equiripple Type 1 five-band linear phase equiripple FIRs with varying error weights.

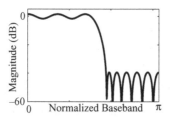 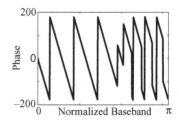

Figure 9.12 Designed 23rd-order equiripple FIR showing magnitude frequency response in decibels (left) and phase response (right).

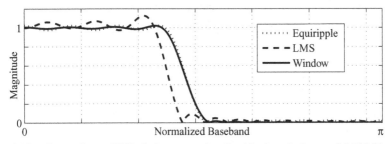

Figure 9.13   Comparison of FIR design strategies. Equiripple, window, and LMS filter magnitude frequency responses.

consider designing and comparing a 31st-order (order 30 MATLAB) window-based, LMS, and equiripple linear phase FIRs that meet a common set of specifications. The window-based filter, designed using a rectangular window, is seen to have an extended transition band and pronounced Gibbs error. The maximum ripple error of the LMS and equiripple filters have essentially the same error found immediately before the end of the passband. It can also be seen that the maximum ripple error of an equiripple filter is constant maximum error, but the LMS local maximum errors monotonically decreases moving toward DC ($\varpi = 0$). As a result, in the absence of any other criterion, any of these may be the designer's selection.

L_p FIR DESIGN

The minimum square error (MSE) and minimax design strategies have been introduced in the form of window-based, LMS, and equiripple designs. These methods give a wide and generally sufficient coverage of the fixed-coefficient FIR design space. A more general design strategy that can embody a number of the studied methodologies is based on minimizing an L_p error norm. The approximation error is again defined to be

$$\varepsilon(\varpi) = W(\varpi)\left|H_d\left(e^{j\varpi}\right) - H\left(e^{j\varpi}\right)\right|, \qquad (9.20)$$

where $W(\varpi) \geq 0$ is a non-negative error weight. It can be noted that the error $\varepsilon(\varpi)$ is the weighted difference between the desired and realized baseband frequency response. The L_p error norm is formally defined to be

$$\left\|\varepsilon(\varpi)\right\|_p = \left(\sum_{i=0}^{N-1}\left(W(\varpi_i)\left|H_d\left(e^{j\varpi_i}\right) - H\left(e^{j\varpi_i}\right)\right|\right)^p\right)^{1/p}. \qquad (9.21)$$

Some of the familiar L_p norms are the $p = 2$ (Euclidian) and $p = \infty$ norms. The L_2 norm satisfies the formula

$$\left\|\varepsilon(\varpi)\right\|_2 = \sqrt{\sum_{i=0}^{N-1}\left(W(\varpi_i)\left|H_d\left(e^{j\varpi_i}\right) - H\left(e^{j\varpi_i}\right)\right|\right)^2}. \qquad (9.22)$$

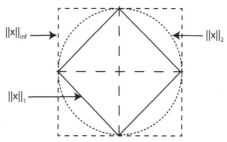

Figure 9.14   Loci of points satisfying $\|\varepsilon(\varpi)\|_p = 1$ for $p = 1$, 2, and ∞.

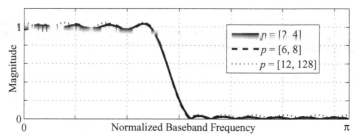

Figure 9.15 L_p design comparison for three ranges of p.

The L_∞ norm satisfies

$$\|\varepsilon(\varpi)\|_\infty = \max\left(W(\varpi_i)\left|H_d\left(e^{j\varpi_i}\right) - H\left(e^{j\varpi_i}\right)\right|\right). \tag{9.23}$$

The differences between L_p norms can be graphically explored. The L_p unit norm (i.e., $\|\varepsilon(\varpi)\|_p = 1$) in two-space is graphically interpreted in Figure 9.14 for $p = 1$, 2, and ∞. While each error norm is unity (i.e., $\|e(\varpi)\|_p = 1$), it can be seen that they produce different outcomes, specifically the domain of $e(\varpi)$ for which $\|e(\varpi)\|_p = 1$.

MATLAB L_p DESIGN

MATLAB provides a means of designing L_p optimal FIRs using the form of firlp-norm, found in the Filter Design Toolbox. The tool can be used to determine the best fit, in an L_p sense, between a desired and realized response. The function firlp-norm performs a scanning operation over a range of p to determine the minimum L_p error design. To illustrate, consider the design of a 21st-order (20 in MATLAB) low-pass filter having a passband out to $f = 0.2f_s$, and three design attempts with the first being $p \in [2, 4]$, the second $p \in [6, 10]$, and the third $p \in [12, 128]$, where p is required to be an even integer with 128 being essentially infinity. The magnitude frequency response outcomes are shown in Figure 9.15. It can be noted that as a group, the designed FIRs are similar. At a more detailed level, however, differences exist. It can be noted that for the lower values of p, the magnitude frequency response

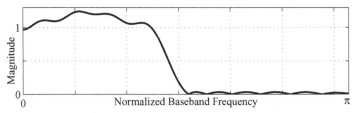

Figure 9.16   Custom L_p FIR design using firlpnorm.

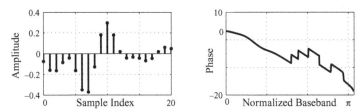

Figure 9.17   Impulse response (left) and phase response (right) of the FIR displayed in Figure 9.16.

takes on the attributes of an LMS FIR filter. For higher values of p, the filter begins to emulate an equiripple design.

One of the features of the L_p design strategy is that firlpnorm is not restricted to model only piecewise constant desired filter frequency responses. To illustrate, suppose the passband of the FIR displayed in Figure 9.15 is to be modified. It is assumed that the middle of the passband is to have a gain that is about 25% higher than the gains found at the passband edges. This is accomplished by using MATLAB to initiate a scan of all even values of p from 2 to 128. The results are shown in Figure 9.16 and exhibit an elevated gain in the middle of the passband as requested.

A word of caution: The resulting filter is not a linear phase FIR as the data in Figure 9.17 would indicate. In particular, the impulse response is asymmetric and the phase response shown to be nonlinear.

CHAPTER 10

FIR: SPECIAL CASES

INTRODUCTION

Finite impulse responses (FIRs) exhibit a wide range of speed, frequency selectivity, and functional capabilities. They are found in a variety of applications. The majority of baseline FIR design requirements can be met using baseline window, least mean square (LMS), or equiripple FIRs. In some instances, these standard design practices can result in excessively complex outcomes or exhibit other undesirable attributes. In such cases, special FIR forms can sometimes be employed to overcome the shortcomings of traditional FIR designs. Some of the more important FIR special cases are developed below.

MOVING AVERAGE (MA) FIR

An important class of FIR filters are those referred to as being multiplier free. A multiplier-free example is the moving average (MA) FIR. In the time domain, the output of an MA filter is the average value of N contiguous samples. An Nth-order MA filter has a transfer function given by

$$H_{MA}(z) = \frac{1}{N}\sum_{m=0}^{N-1} z^{-m} = \frac{1}{N}\sum_{m=0}^{\infty} z^{-m} - \frac{1}{N}\sum_{m=N}^{\infty} z^{-m} = \frac{1}{N}\left(\frac{1}{1-z^{-1}}\right) - \frac{1}{N}\left(\frac{z^{-N}}{1-z^{-1}}\right)$$

$$= \frac{1}{N}\left(\frac{1-z^{-N}}{1-z^{-1}}\right) = \frac{1}{N}\frac{\left(z^{N}-1\right)}{z^{N-1}\left(z-1\right)}. \tag{10.1}$$

An Nth-order MA FIR is implemented with $N-1$ shift registers and an accumulator. The accumulator's dynamic range requirement can be specified in terms of the filter's worst case gain. The MA FIR's worst case gain is given by $G_{MA} = (1/N)\sum |h[k]|$. Since the FIR's impulse response is $h[k] = 1$ at sample k, the worst case gain is given by $G_{MA} = 1$. The zeros or nulls of the MA FIR are defined by the location of the roots of the numerator polynomial in Equation 10.1, namely,

$$z^{N} - 1 = 0 \rightarrow z_i = e^{(j2\pi i/N)} \text{ for } i \in [0, N-1]. \tag{10.2}$$

Digital Filters: Principles and Applications with MATLAB, First Edition. Fred J. Taylor.
© 2012 by the Institute of Electrical and Electronics Engineers, Inc.
Published 2012 by John Wiley & Sons, Inc.

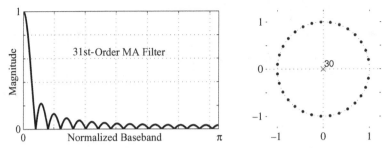

Figure 10.1 Thirty-first-order moving average (MA) multiplier-free FIR. Magnitude frequency response (left) and pole-zero distribution (right). Note the absence of a pole or zero at $z = 1$.

The zeros of an Nth-order MA filter are uniformly distributed about the periphery of the unit circle ($z = e^{j2\pi n/N}$) on $\phi = 2n/N$ radian centers. One of the zeros is located at $z = 1$, which exactly coincides with the location of a pole defined by the denominator found in Equation 10.1. The result is exact pole-zero cancellation at $z = 1$ as displayed in Figure 10.1. The surviving zeros define the location of the frequency response nulls shown in Figure 10.1. The MA FIR's magnitude frequency response is seen to be decidedly low-pass and has a $\sin(x)/x$ magnitude frequency response envelope.

COMB FIR

A comb FIR is a multiplier-free variation on the MA filter theme. An Nth-order comb FIR simply adds or subtracts an N-sampled delayed from the current input sample value. The two versions of a comb FIR have transfer functions given by

$$H_+(z) = 1 + z^{-N} = 0 \rightarrow z_i = e^{(j2\pi i/N + j\pi)} \text{ for } i \in [0, N-1],$$
$$H_-(z) = 1 - z^{-N} = 0 \rightarrow z_i = e^{(j2\pi i/N)} \text{ for } i \in [0, N-1].$$

(10.3)

Equation 10.3 states that a multiplier-free comb filter consists only of shift registers and an adder or subtractor. Like the MA filter, the zeros of the comb filter $H_\pm(z)$ are uniformly distributed along the periphery of the unit circle. The locations of the zeros define the position of the nulls in the frequency domain as shown in Figure 10.2. Observe that for the (+) case, the comb filter does not have a null at 0 Hz (DC) ($z = 1$), resulting in a finite DC gain (i.e., $H_+(1) = 2$). For the (−) case, the comb filter has a null located at DC ($z = 1$) and has local peak gains of 2 between the nulls. In either case, the FIR's worst case comb filter gain is $G_{comb} = 2$.

Example: MA and Comb Zero Locations

The pole-zero distributions of unscaled MA and comb filters, both having transfer function numerators ($1 \pm z^{-7}$), are displayed in Figure 10.3. Observe that in all cases the zeros lie on the unit circle and are separated from each other by $2\pi/7$ radians.

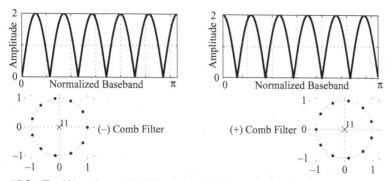

Figure 10.2 Twelfth-order comb FIR magnitude frequency response and zero distribution for the (–) option (left) and (+) option (right).

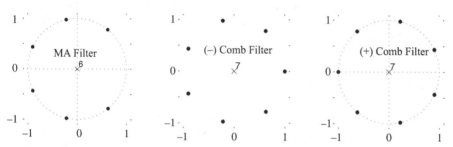

Figure 10.3 Pole-zero distributions of moving average and comb FIR filters all having numerators of $(1 - z^{-7})$. Notice that the zeros are separated by angle $2\pi/7$ radians.

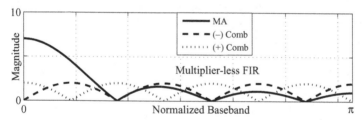

Figure 10.4 Magnitude frequency responses for an unnormalized MA FIR ($h_1[k] = \{1, 1, 1, 1, 1, 1, 1, 1\}$) ($N = 8$), a (–) comb filter ($h_2[k] = \{1, 0, 0, 0, 0, 0, 0, -1\}$) ($N = 7$), and (+) comb filter ($h_3[k] = \{1, 0, 0, 0, 0, 0, 0, 1\}$) ($N = 7$).

The MA pole-zero annihilation occurs at $z = 1$ giving rise to a finite DC gain (i.e., $H(1) \neq 0$). The comb filter, given by $H_+(z) = 1 + z^{-7}$, also has a finite DC gain by virtue of the fact that the filter does not possess a zero $z = 1$. However, the filter has zero gain at the Nyquist frequency due to a zero being located at $z = -1$. The comb filter given by $H_-(z) = 1 - z^{-7}$ has a zero at $z = 1$, but none at $z = -1$. Therefore, this filter has a zero DC gain and finite gain at the Nyquist frequency. The worst case gain for the unnormalized MA filter is $G_{MA} = 8$, and for the comb filters, $G_{comb} = 2$. The magnitude frequency responses for the three multiplier-free FIRs are examined in Figure 10.4.

L-BAND FILTERS

An *L*-band *N*th-order FIR is a filter with every *L*th coefficient being zero. An *L*-band low-pass FIR is also called a Nyquist filter and has a passband width that is roughly 1/*L*th the baseband frequency range. Compared with a regular *N*th-order FIR defined in terms of *N* nonzero tap-weight coefficients, an *L*-band *N*th-order FIR is less complex. The fact that every *L*th multiply-accumulate can be omitted means that an *L*-band FIR has potentially higher real-time bandwidth when compared with a general *N*th-order FIR.

A half-band FIR ($L = 2$) is a special FIR case that exhibits the magnitude frequency response similar to that shown in Figure 10.5. Observe that the frequency response has a point of symmetry in the middle of the baseband (i.e., at $\omega = \pi/2$), a point that corresponds to half the Nyquist frequency. The result is an essential 2:1 reduction in the number of coefficients needed to implement a half-band FIR compared with the general case. That is, except for the center-tap coefficient, every other tap-weight coefficient is zero, effectively halving the FIR's multiply-accumulation burden.

Example: Half-Band FIR

Consider an 11th-order linear phase half-band FIR having the center of the transition band set to $f = 0.25f_s$ with a transition bandwidth of $\Delta f = \pm 0.05f_s$. The filter coefficients of the resulting half-band FIR filter are shown in Figure 10.6. Notice that except for the FIR's center-tap coefficient, every other coefficient has a value of zero. As a result, when compared with an arbitrary FIR, the half-band FIR requires approximately ½ the number of multiply-accumulate calls per filter cycle. This translates into a higher potential real-time filter rate.

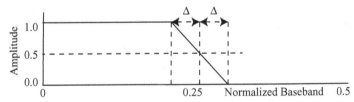

Figure 10.5   Desired magnitude frequency response of a half-band FIR.

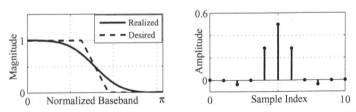

Figure 10.6   Magnitude frequency response of a low-pass 11th-order half-band FIR (left) with a symmetric impulse response (right) showing FIR coefficients $h[1]$, $h[3]$, $h[7]$, and $h[9] = 0$.

MIRROR FIR

FIR windowed, LMS, and equiripple strategies are core FIR technologies. In all cases, a noncausal transfer function of the resulting Lth-order FIR has the form $H(z) = \Sigma\ h[k]z^{-k}$, $k \in [-M, M]$, where $L = 2M + 1$. It is often desired to develop variations of these basic filters that maintain a close mathematical relationship to the original filter, but physically exhibits different frequency-domain behavior. For example, a mirror filter reflects an inverted copy of the magnitude frequency response of a parent filter. If the original, or parent filter, is low-pass with a normalized transition bandwidth Δ, then the mirror filter would be a high-pass filter having the same transition bandwidth Δ. Mirroring can be achieved by simply modulating the impulse response $h[k]$ by a sinusoid running at the Nyquist frequency. That is, the parent filter's coefficients are modulated by $c[k] = \cos(\pi k) = (-1)^k$. This action will translate (heterodyne) the original frequency response centered about DC to a new center frequency located at the Nyquist frequency. As a result, the mirror version of a parent FIR $h_{\text{parent}}[k]$ has an impulse response $h_{\text{mirror}}[k] = (-1)^k h_{\text{parent}}[k]$. The only structural difference between the mirror and parent filter is the alternating sign of the tap-weight coefficients. The spectral relationship between a mirror and its parent filter is graphically interpreted in Figure 10.7 for a linear phase 21st-order parent FIR.

COMPLEMENT FIR

Assume the transfer function of a parent filter is given by $H_{\text{parent}}(e^{j\omega})$ filter and has an impulse response $h_{\text{parent}}[k]$ over $k \in [-L, L]$. The transfer function of the complementary filter is

$$H_{\text{comp}}(e^{j\omega}) + H_{\text{parent}}(e^{j\omega}) = 1. \qquad (10.4)$$

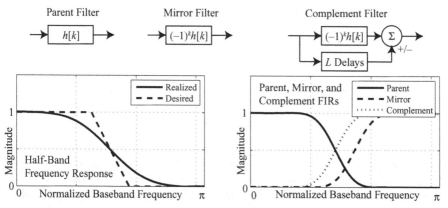

Figure 10.7 Comparison of mirror and complement (see next section) FIRs with a common low-pass parent FIR.

If the odd order linear phase Type 1 parent filter $H_{\text{parent}}(e^{j\varpi})$ has a passband gain of K, then Equation 10.4 can be modified to read $H_{\text{comp}}(e^{j\varpi}) + H_{\text{parent}}(e^{j\varpi}) = K$. Observe that when the original and complement filter are added, an all-pass filter results. Complement filters can be particularly useful when designing a bank of sub-band filters to cover a wide frequency range. From Equation 10.4, the complement filter can be expressed as (assume $K = 1$) $H_{\text{comp}}(e^{j\varpi}) = 1 - H(e^{j\varpi})$ with an impulse response $h_{\text{comp}}[k]$, over $k \in [-L, L]$. To create and implement a complement filter, the center-tap coefficient (i.e., $h[0]$) of the parent FIR is subtracted from unity to define a new center-tap coefficient for the complement FIR. The other complement filter coefficients are the negative of the parent filter's tap-weight values. If the parent's center-tap coefficient is not located at $k = 0$ but is instead at sample L for a causal filter, then the parent and complement FIR's impulse responses are given by

$$\text{Parent: } h[k] = \{h[0], \ldots, h[L-1], h[L], h[L+1], \ldots, h[7L]\}$$
$$\text{Complement: } h_{\text{comp}}[k] = \{-h[0], \ldots, -h[L-1], (1 - h[L]), -h[L+1], \ldots, -h[2L]\}.$$

$$(10.5)$$

In terms of Equation 10.5, a complement FIR's transfer function can be interpreted as $H_{\text{comp}}(z) = z^{-L} - H_{\text{parent}}(z)$, which states that the complement filter can be obtained simply with the addition of the output of an L-delay shift register, which is a multiplier-less operation (essentially zero overhead). It is interesting to note that the shift register length L is equal to the group delay of the linear phase parent FIR. This case is illustrated in Figure 10.7.

Example: Mirror and Complement FIR

Consider a simple linear phase fifth-order unnormalized parent MA parent FIR having an impulse response $h[k] = \{1/5, 1/5, 1/5, 1/5, 1/5\}$. The mirror and complement versions are given by

$$h_{\text{mirror}}[k] = \{1/5, -1/5, 1/5, -1/5, 1/5\} \Leftrightarrow H_{\text{mirror}}(z)$$
$$= 1/5 - 1/5z^{-1} + 1/5z^{-2} - 1/5z^{-3} + 1/5z^{-4}$$

and

$$h_{\text{comp}}[k] = \{-1/5, -1/5, 4/5, -1/5, -1/5\} \Leftrightarrow H_{\text{mirror}}(z)$$
$$= -1/5 - 1/5z^{-1} + 4/5z^{-2} - 1/5z^{-3} - 1/5z^{-4}.$$

The magnitude frequency responses of the fifth-order parent MA, mirror, and complement filters are displayed in Figure 10.8. The mirror action is obvious and the complement filter behavior can be argued in terms of $H_{\text{comp}}(e^{j\varpi}) + H_{\text{parent}}(e^{j\varpi}) = 1$. Specifically, $H_{\text{comp}}(e^{j\varpi}) + H(e^{j\varpi}) = z^{-2}$, which physically represents an all-pass linear phase FIR.

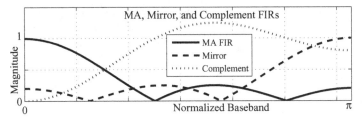

Figure 10.8 Fifth-order MA parent FIR, mirror, and complement versions.

FREQUENCY SAMPLING FILTER BANK

The frequency sampling filter bank consists of a set of band-pass filters. It draws from the theory of multiplier-free FIRs and should not be confused with a frequency sampling filter based on the theory of truncated Fourier transform (MATLAB fir2). An Nth-order frequency sampling filter bank can be envisioned as a collection of narrow-band filters having user-defined gains ($A(\varpi)$) and phase responses ($\phi(\varpi)$) at the normalized center frequencies $\varpi = 2\pi n/N$. That is, at the nth center frequency, the filter's response is

$$H_n(e^{j2\pi n/N}) = H_n[n] = A(n)\angle\phi(n). \qquad (10.6)$$

The impulse response of an Nth-order FIR satisfying Equation 10.6, for $n \in [0, N-1]$, can be computed using an inverse discrete Fourier transform (IDFT). Equivalently, the z-transform and DFT representation of the realized FIR is

$$H(z) = \sum_{k=0}^{N-1} h[k]z^{-k} = \frac{1}{N}\sum_{k=0}^{N-1}\left(\sum_{n=0}^{N-1} H_n[n]e^{j2\pi nk/N}\right)z^{-k}, \qquad (10.7)$$

which, after simplification and reversing the order of summation, can be further reduced to read

$$H(z) = \frac{1}{N}\sum_{n=0}^{N-1} H_n[n]\left(\sum_{k=0}^{N-1}\left(e^{j2\pi n/N}z^{-1}\right)^k\right). \qquad (10.8)$$

Using a standard summation reduction formula, the parenthetical term in Equation 10.8 can be expressed as

$$\sum_{k=0}^{N-1}\left(e^{j2\pi n/N}z^{-1}\right)^k = \frac{1-z^{-N}}{1-e^{j2\pi n/N}z^{-1}}. \qquad (10.9)$$

This allows the FIR transfer function to be expressed as

$$H(z) = \frac{(1-z^{-N})}{N}\sum_{n=0}^{N-1}\left(\frac{H_d[n]}{1-e^{j2\pi n/N}z^{-1}}\right) \qquad (10.10)$$

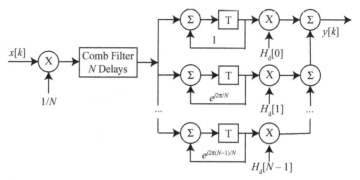

Figure 10.9   Frequency sampling filter bank system architecture.

and is interpreted in Figure 10.9. The preamble filter $(1 - z^{-N})$ is seen to be a simple comb filter that places nulls (zeros) at the reference frequency locations $\omega = 2\pi n/N$. The terms in the summation represent a collection of sub-band filters having poles also at the center frequencies $\varpi = 2\pi n/N$. Ideally, the pole at $\varpi = 2\pi n/N$ cancels a comb filter zero located at the same location. Due to the pole-zero cancellation, the resulting filter is without poles (i.e., FIR).

Each sub-band filter is a sharply tuned narrow-band filter characterized by

$$H_n(z) = \frac{H_d[n]}{1 - e^{j2\pi n/N}z^{-1}} = \frac{A[n]\angle\phi(n)}{1 - e^{j2\pi n/N}z^{-1}}, n \in [0, N-1]. \tag{10.11}$$

The poles of these resonator filters are located along the periphery of the unit circle at $z = e^{j2\pi n/N}$. For stability reasons, the filter poles and zeros are often moved slightly interior to the unit circle by scaling the unit circle poles and zeros back to a radius r (i.e., $z = re^{j2\pi n/N}$), where $r < 1$. Since the complex poles occur in complex-conjugate pairs that allow the filter defined by Equation 10.10 to be represented as a second-order system (using Euler's equation), namely,

$$H(z) = \frac{(1 - r^N z^{-N})}{N}\left(\frac{H_n[0]}{1 - rz^{-1}} + \sum_{n=1}^{(N-1)/2} \frac{2A[n](\cos(\varphi(n)) - (r\cos(\varphi(n) - 2\pi n/N)z^{-1})}{1 - 2r\cos(2\pi n/N)z^{-1} + r^2 z^{-2}}\right)$$

$$\tag{10.12}$$

if N is odd. If N is even, then

$$H(z) = \frac{(1 - r^N z^{-N})}{N} \times$$

$$\left(\frac{H_d[0]}{1 - rz^{-1}} + \frac{H_d[N-1]}{1 + rz^{-1}} + \sum_{n=1}^{N/2-1} \frac{2A[n](\cos(\varphi(n)) - (r\cos(\varphi(n) - 2\pi n/N)z^{-1})}{1 - 2r\cos(2\pi n/N)z^{-1} + r^2 z^{-2}}\right).$$

$$\tag{10.13}$$

While the frequency sampling filter bank method has been successfully used for some time to design FIR filters having an arbitrary magnitude frequency response envelope, it does have acknowledged limitations. A major design problem can occur when a frequency sampling filter is required to be modeled as an ideal filter having a piecewise constant filter response characterized by $|H(n)| = 1$ and $|H(n + 1)| = 0$ (i.e., abrupt change across the transition band boundary). Modeling sharp transition is also difficult using an IDFT (i.e., Gibbs phenomenon). The problem can be mitigated somewhat by relaxing the transition band condition so that $|H(n)| = 1 - \alpha$, and $|H(n + 1)| = \alpha$ for $0 \angle \alpha \angle 1.0$. This will reduce the slope of the filter's skirt so that it can be more easily modeled by an IDFT.

Example: Frequency Sampling Filter

An $N = 15$ frequency sampling band-pass FIR having an assigned frequency response,

$$H_n(n) = \begin{cases} 1 + j0, n = \pm 3 \\ 0 + j0, \text{otherwise} \end{cases},$$

is to be implemented. The solution consists of four filters: $H_0(z)$, $H_1(z)$, $H_2(z)$, and $H_3(z)$. The frequency assignment for $H_3(z)$ ($n = 3$), at frequency $\varpi = 2\pi(3/15)$, satisfies

$$H_3(z) = \frac{(1 - z^{-15})}{15} \frac{2(1 - \cos(2\pi(3/15))z^{-1})}{1 - 2\cos(2\pi(3/15))z^{-1} + z^{-2}}.$$

The resulting frequency sampled filter's magnitude frequency responses and pole-zero distribution is shown in Figure 10.10. The frequency response is seen to be that of a band-pass filter. The pole-zero diagram shows pole-zero cancellation at $\varpi = \pm 2\pi(3/15)$, leaving all other surviving zeros unaffected.

SAVITZKY–GOLAY (SG) FIR

SG smoothing filters (also called digital smoothing polynomial filters) are typically used to reduce the effects of additive noise in a broadband signal. They are not so much a filter as they are a polynomial interpolator. In some applications, SG

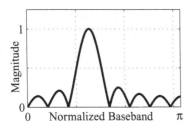
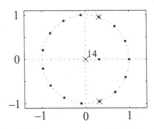

Figure 10.10 Magnitude frequency response of a single frequency sampling filter bank band-pass filter element (left) and pole-zero distribution (right).

smoothing filters can outperform a standard FIR smoothing filter, which tends to filter out a significant portion of the signal's high-frequency content along with the noise. Although SG filters are more effective at preserving pertinent high-frequency signal components, they are less successful than standard averaging FIR filters at rejecting low-frequency noise. The response of an SG filter to an input $x[k]$ is

$$y[i] = \sum_{k=-n_L}^{n_R} h_0[k] x[i+k]. \tag{10.14}$$

The SG smoothing strategy is designed to preserve higher-order statistical moments and is based on least-squares techniques that fit a low-order polynomial (typically quadratic or quadric) to a collection of data samples. Data are assumed to belong to a moving noisy signal record of length $N_R + N_L + 1$ samples (N_R = number of samples to the right of the 0th sample, N_L to the left). A polynomial interpolation equation can then be defined by

$$\hat{x}[d] = \sum_{n=-m}^{m} a_m d^n, \tag{10.15}$$

where $\hat{x}[d]$ is to become an estimate of $x[d]$, for $d \in [-N_L, N_R]$. An example of a quadratic interpolation filter is given by $\hat{x}[d] = a_0 1 + a_1 d + a_2 d^2$. The LMS solution can be expressed in terms of the matrix equation

$$Da = \hat{x}; \; D = \begin{bmatrix} d_a^0 & d_a^1 & \dots & d_a^n \\ d_b^0 & d_b^1 & \dots & d_b^n \\ \dots & \dots & \dots & \dots \\ d_q^0 & d_q^1 & \dots & d_q^n \end{bmatrix}, \tag{10.16}$$

where d_k is the relative distance from the point to be smoothed to data point $\hat{x}[k]$. The optimal solution will minimize the squared error where the error is given by $\varepsilon = x - \hat{x} = x - D\hat{x}$ with the solution that is defined by the pseudoinverse (or Penrose inverse), namely,

$$a = (D^T D)^{-1} D^T x = Hx. \tag{10.17}$$

For example, using a 0th-order interpolating polynomial, namely, $\hat{x}[k] = a_0$, and five consecutive samples $\{x_{-2}, x_{-1}, x_0, x_1, x_2\}$, with $N_L = N_R = 2$, define the interpolator $D = [1\,1\,1\,1\,1]^T$. Since $D = [1\,1\,1\,1\,1]^T$, it follows that

$$H = (D^T D)^{-1} D^T = \frac{1}{5} \begin{bmatrix} 1 & 1 & 1 & 1 & 1 \\ 1 & 1 & 1 & 1 & 1 \\ 1 & 1 & 1 & 1 & 1 \\ 1 & 1 & 1 & 1 & 1 \\ 1 & 1 & 1 & 1 & 1 \end{bmatrix}. \tag{10.18}$$

Isolating a_0, it follows that $a_0 = (1/5)[1\,1\,1\,1\,1]^T$. Finally,

$$\hat{x}_0 = a_0 x = \frac{1}{5}\sum_{k=-2}^{2} x_k, \qquad (10.19)$$

which defines the synthesized filter to be a simple MA estimator. The process can be generalized to read

$$a = \left(D^T D\right) D^T x$$

$$y_0[n] = a_0^T x = \sum_{m=-N_L}^{N_R} a_0[m]\,x[n+m] \qquad (10.20)$$

where $y_0[n]$ is the optimal noise-suppressed estimate of the interpolated value of a set of noisy sample values.

Example: SG FIR Design

Use a cubic interpolating polynomial having the form $a_0 + a_1 d + a_2 d^2 + a_3 d^3$, $d \in [-2, 2]$, and knowledge of five consecutive samples, with $N_L = 2$ and $N_R = 2$, define an SG smoothing filter in terms of

$$D = \begin{bmatrix} 1 & -2 & (-2)^2 & (-2)^3 \\ 1 & -1 & (-1)^2 & (-1)^3 \\ 1 & 0 & 0 & 0 \\ 1 & 1 & (1)^2 & (1)^3 \\ 1 & 2 & (2)^2 & (2)^3 \end{bmatrix} = \begin{bmatrix} 1 & -2 & 4 & -8 \\ 1 & -1 & 1 & -1 \\ 1 & 0 & 0 & 0 \\ 1 & 1 & 1 & 1 \\ 1 & 2 & 4 & 8 \end{bmatrix};$$

$$(D^T D)^{-1} D^T = \frac{1}{35}\begin{bmatrix} 34.5 & 2 & -3 & 2 & -05 \\ 2 & 27 & 12 & -8 & 2 \\ -3 & 12 & 17 & 12 & -3 \\ 2 & -8 & 12 & 27 & 2 \\ -0.5 & 2 & -3 & 2 & 34.5 \end{bmatrix}.$$

SG filters are optimal in that they minimize the least-square error in fitting a polynomial to each frame of noisy data. An example of an SG application is suppressing additive noise from a noisy sinusoid $x[k]$, as shown in Figure 10.11. The displayed outcome was produced using MATLAB's `sgola(x, k, f)` command, where k is the interpolation polynomial order (e.g., cubic = 3) and f is called the frame size (active data window). The results shown in Figure 10.11 are for a cubic interpolation ($k = 3$) and two frame sizes ($f = 31, 101$).

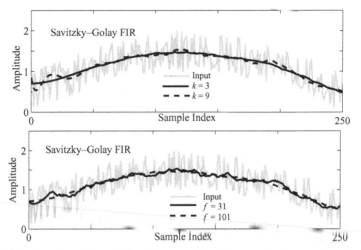

Figure 10.11   Savitzky–Golay (SG) cubic smoothing filters applied to a noise-added sinusoid using different frame sizes.

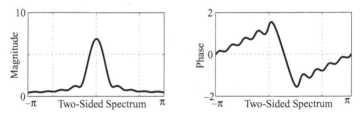

Figure 10.12   Magnitude frequency response and phase response of an 11th-order nonlinear phase FIR.

NONLINEAR PHASE FIR

An acknowledged FIR attribute is its ability to support linear phase signal processing. There are times, however, when maintaining phase linearity is not required, but minimizing filter order (complexity) is an objective. In such cases, a nonlinear phase FIR becomes a viable design option. Due to symmetry requirements, the design specifications of an Nth-order linear phase FIR essentially constrains N coefficients ($h_i = \pm h_{N-i}$). Without such a constraint, an FIR of approximate order $N/2$ may be found that meets or exceeds the magnitude frequency specification (no imposed phase requirements). The MATLAB function cfirm program can be used to implement a nonlinear phase equiripple FIR.

Example: Nonlinear Phase FIR

Design an 11th-order nonlinear phase FIR with an asymmetric impulse response given by the raised cosine function $1 + \cos(\phi)$, $\phi \in [0, \pi]$. The magnitude frequency response and phase response are displayed in Figure 10.12. To achieve the displayed

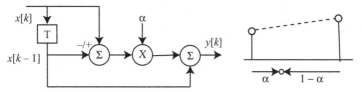

Figure 10.13   Farrow filter using linear interpolation.

frequency response, a 22nd-order linear phase FIR would be required. The nonlinear phase performance can be witnessed in Figure 10.12.

FARROW FIR

An interesting class of filter is one that can implement fractional delays to a signal with respect to the filter's clock. Such a filter can be motivated in the context of interpolation. Interpolation, it may be recalled, refers to those mathematical techniques that produce intersample values of a sampled signal process. A system that can implement variable delay lengths interpolations is called a Farrow filter. A Farrow filter based on a linear interpolation rule is shown in Figure 10.13. The system described satisfies the difference equation:

$$\text{Fractional delay: } y[k] = (1-\alpha)x[k-1] + \alpha x[k]. \tag{10.21}$$

The choice of the control parameter α will allow the interpolation delay to be effectively set between 0 and 1.

CHAPTER 11

FIR IMPLEMENTATION

FINITE IMPULSE RESPONSE FILTER (FIR) IMPLEMENTATION

Digital filters are implemented using the basic building block elements of adders, multipliers, and shift registers. How these elements are arranged and interconnected defines a filter's architecture. In general, a given filter can have multiple architectures that can be used to implement a common transfer function. The individual architectures possess different attributes in the form of complexity, speed, latency, accuracy, power dissipation, and other metrics. There are, however, only a few basic architectural forms that are found in common use.

DIRECT-FORM FIR

An Nth-order causal FIR filter, having an impulse response given by $h[k] = \{h_0, h_1, \ldots, h_{N-1}\}$, can be expressed in transfer function form as $H(z) = \sum h_k z^{-k}$, $k \in [0, N-1]$. The most common FIR implementation architecture is called the direct-form FIR. A direct FIR implementation of $H(z)$ is summarized in Figure 11.1. An Nth-order direct FIR is seen to consist of a collection of $N-1$ shift registers, N tap-weight coefficients h_k, with attendant multipliers, and $N-1$ adders or an accumulator. The FIR's impulse response can be directly inferred from the architecture to be $h[k] = \{h_0, h_1, \ldots, h_{N-1}\}$. For each input sample $x[k]$, a direct FIR would be implemented using the following set of arithmetic operations:

For each input sample $x[k]$, do

$$x_0 = x[k]$$
$$y[k] = h_0 x_0 + h_1 x_1 + \ldots + h_{N-1} x_{N-}$$
$$\{\text{update FIFO stack}; x_{N-1} = x_{N-2}, \ldots = \ldots, x_2 = x_1, x_1 = x_0\}$$

Once initiated, the routine would be continually executed over the filter's life cycle. The designers of modern digital signal processing (DSP) μps have learned to implement arrayed multiply-accumulate (MAC) calls using optimized internal architectures involving single or multiple MAC units. Modern DSP μps generally contain dual-port memory that can simultaneously supply two MAC operands (x_i, h_j) per

Digital Filters: Principles and Applications with MATLAB, First Edition. Fred J. Taylor.
© 2012 by the Institute of Electrical and Electronics Engineers, Inc.
Published 2012 by John Wiley & Sons, Inc.

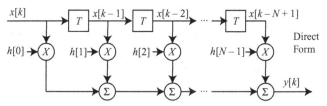

Figure 11.1 Direct Nth-order FIR architecture.

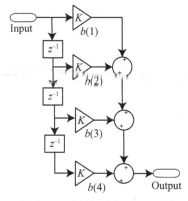

Figure 11.2 A fourth-order (third-order MATLAB) direct FIR architecture (MATLAB style).

execution cycle. Coefficients can be read to and from memory in a circular modulo(N) manner, repeating a fixed sequence for each filter cycle. Since many DSP μps can execute an instruction in a single MAC or pipeline cycle, the computational latency can be estimated to be $T_{\text{FIR\_cycle}} \cong (N + 1)T_{\text{inst-cycle}}$

Mathworks provides some direct architectural support for implementing FIRs using the MATLAB Filter Design Toolbox dfilt option. In particular, dfilt.dffr converts a transfer function into a direct FIR. To illustrate, a fourth-order FIR is presented in Figure 11.2. In addition, MATLAB's Signal Processing Toolbox contains the fvtool command that can be used to graphically examine the filter's time-domain, pole-zero, and frequency-domain behavior. A sample of the MATLAB code is shown below:

```
h=firpm(3,[0 0.4 0.5 1], [1 1 0 0]);  % Filter Design Toolbox

Hdir=dfilt.dffir(h);  % Direct FIR

realizemdl(Hdir);  % SIMULINK
```

TRANSPOSE ARCHITECTURE

Another baseline FIR architecture is called the transpose FIR, which is a variation of the direct architecture theme. An FIR, with an impulse response $h[k] =$

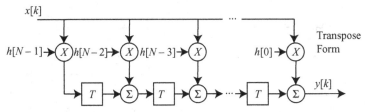

$x[k]$

$h[N-1] \rightarrow$ (X) $h[N-2] \rightarrow$ (X) $h[N-3] \rightarrow$ (X) $h[0] \rightarrow$ (X)   Transpose Form

$y[k]$

Figure 11.3   Transpose Nth-order FIR architecture.

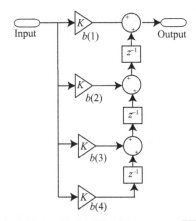

Figure 11.4   A fourth-order (third-order MATLAB) transpose FIR architecture (MATLAB style).

$\{h_0, h_1, \ldots, h_{N-1}\}$ can be implemented as the transpose architecture shown in Figure 11.3. Mathworks provides some support in the creation of a transpose FIR using the Filter Design Toolbox dfilt option. In particular, dfilt.dffirt converts an FIR's transfer function into a transpose form FIR. The outcome for a fourth-order FIR is shown in Figure 11.4. A sample of the MATLAB code is shown below:

```
h=firpm(3,[0 0.4 0.5 1], [1 1 0 0]);  %  Filter Design Toolbox

Hdt=dfilt.dffirt(h); % Transpose FIR

realizemdl(Hdt); % SIMULINK
```

SYMMETRIC FIR ARCHITECTURES

Many baseline FIRs are linear phase filters and, as a result, possess either even or odd coefficient symmetry. Coefficient symmetry allows the direct-form FIR architecture to be modified as shown in Figure 11.5, resulting in what is called a symmetric FIR architecture. The symmetric filter's advantage is a reduced multiplier budget. If a filter has an add/subtract cycle time that is less than a multiply cycle

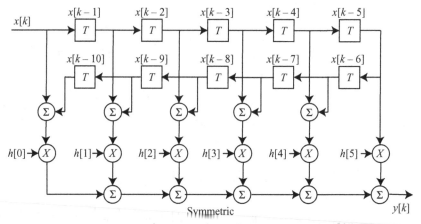

Figure 11.5   Reduced multiplication symmetric FIR architecture for an 11th-order FIR.

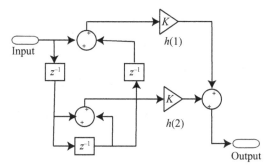

Figure 11.6   A fourth-order (third-order MATLAB) symmetric FIR architecture (MATLAB style).

time, then a symmetric FIR can achieve a higher real-time bandwidth. This presumes that adders and multipliers are always available when needed. A conventional DSP μp, however, has been optimized to perform a series of in-order MACs and not add-multiply-add operations required of symmetric FIR. As a result, symmetric FIR architectures are not synergist with general-purpose DSP μp architecture.

Mathworks provides some support in the creation of a symmetric FIR using the Filter Design Toolbox dfilt option. In particular, dfilt.dfsymfir converts a transfer function into a symmetric form FIR of even or odd order. The outcome for a fourth-order FIR is shown in Figure 11.6. A sample of the MATLAB code is shown below:

```
h=firpm(3,[0 0.4 0.5 1], [1 1 0 0]); %  Filter Design Toolbox

Hsym=dfilt.dfsymfir(h); % Symmetric FIR

realizemdl(Hsym); % SIMULINK
```

LATTICE FIR ARCHITECTURE

Another important FIR structure is called the lattice architecture. An Nth-order lattice FIR is shown in Figure 11.7. The lattice coefficient $k[i]$ is called a partial correlation (PARCOR) coefficient, which is related to the FIR's impulse response. A lattice FIR requires $2N$ multiples per filter cycle (vs. N for a direct FIR) and therefore is more complex. Nevertheless, a lattice FIR is often preferred over the direct FIR due to a natural ability to suppress coefficient round-off errors. In addition, lattice structures are often the architecture of choice when implementing adaptive filters and applications that do not require linear phase behavior. A lattice architecture can be used to implement a monic FIR filter having the general form

$$A(z) = \left[1.0 + \sum_{j=1}^{N-1} a_j z^{-j} \right],$$ (11.1)

where the use of the coefficient notation is purposeful. Specifically, the coefficient a_j is differentiated from the coefficient $k[i]$ found in Figure 11.7. The filter described by Equation 11.1 can also be implemented using a direct or transpose FIR architecture. Due to the structure of $A(z)$, it can also be seen that a lattice FIR has asymmetric impulse response, precluding its ability to implement a linear phase filter. To fully appreciate the lattice architecture, consider the simple second-order lattice filter

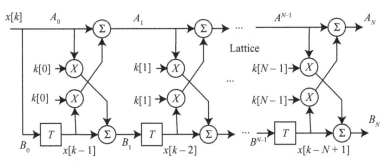

Figure 11.7   Lattice FIR architecture.

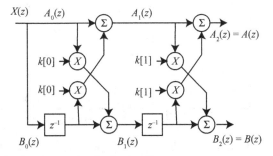

Figure 11.8   Motivational second-order lattice filter.

shown in Figure 11.8 in the z-transform domain. The input–output relationships can be expressed as

$$A(z) = \left(1 + k[0]z^{-1} + k[1]z^{-2}\right)X(z),$$
$$B(z) = \left(k[1] + k[0]z^{-1} + z^{-2}\right)X(z). \tag{11.2}$$

It can be seen that $A(z)$ carries the form suggested in Equation 11.1. Furthermore, $B(z)$ is $A(z)$ in permuted (reversed) order.

The coefficients shown in Equation 11.1 are related to the coefficients of the lattice filter of Figure 11.7 over iteration indices $i = 1, 2, \ldots, N - 1$. Specifically, let $A_0(z) = B_0(z) = 1.0$. The PARCOR coefficient $k[i]$, shown in Figure 11.7, is related to a_j of Equation 11.1, as follows:

$$A_m(z) = A_{m-1}(z) + k_m z^{-1} D_{m-1}(z); \; m \in [1, N-1]$$
$$B_m(z) = z^{-m} A_m(z^{-1}); \; m \in [1, N-1] \tag{11.3}$$

where $B_m(z)$ is simply $A_m(z)$ in reverse order. In reversing the process, it follows that

$$A_{m-1}(z) = \frac{A_m(z) - k_m A_m(z)}{1 - k_m^2}; \; m \in [1, N-1]. \tag{11.4}$$

Example: Lattice FIR

Suppose a third-order lattice filter, shown in Figure 11.9, has PARCOR coefficients $k_1 = 1/4$, $k_2 = 1/2$, and $k_3 = 1/3$. The direct architecture coefficient a_i can be computed as follows:

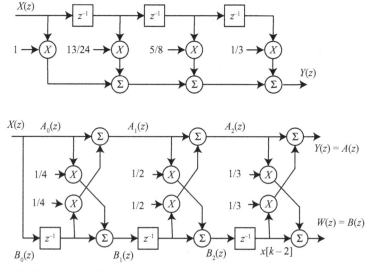

Figure 11.9 Third-order FIR with a direct architecture (top) and lattice filter architecture (bottom).

m	$A_m(z)$	$B_m(z)$
1	$A_1(z) = A_0(z) + k_1 z^{-1} B_0(z) = 1 + k_1 z^{-1} = 1 + \dfrac{1}{4} z^{-1}$	$B_1(z) = z^{-1} A_1(z^{-1}) = \dfrac{1}{4} + z^{-1}$
2	$A_2(z) = A_1(z) + k_2 z^{-1} B_1(z) = 1 + \dfrac{1}{4} z^{-1} + \dfrac{1}{2} z^{-1}\left(\dfrac{1}{4} + z^{-1}\right)$ $= 1 + \dfrac{3}{8} z^{-1} + \dfrac{1}{2} z^{-2}$	$B_2(z) = z^{-2} A_2(z^{-1}) = \dfrac{1}{2} + \dfrac{3}{8} z^{-1} + z^{-1}$
3	$A_3(z) = A_2(z) + k_3 z^{-1} B_2(z) = 1 + \dfrac{3}{8} z^{-1} + \dfrac{1}{2} z^{-2} +$ $\dfrac{1}{3} z^{-2}\left(\dfrac{1}{2} + \dfrac{3}{8} z^{-1} + z^{-1}\right) = 1 + \dfrac{13}{24} z^{-1} + \dfrac{5}{8} z^{-2} + \dfrac{1}{3} z^{-2}$	$B_3(z) =$ not required

The resulting direct filter coefficients are $a_0 = 1$, $a_1 = 13/24$, $a_2 = 5/8$, and $a_3 = 1/3$. Finally, the transfer function defined from input to the output $Y(z)$ is $A_3(z) = A(z)$ and $W(z) = B_3(z) = B(z)$. As a side note, the polynomials $A_1(z)$, $B_1(z)$, $A_2(z)$, $B_2(z)$, $A_3(z)$, and $B_4(z)$ correspond to the transfer function from input to their corresponding output locations.

Reversing the process, one obtains the following:

$A_m(z)$	k	$B_m(z)$
$A_3(z) = 1 + \dfrac{13}{24} z^{-1} + \dfrac{5}{8} z^{-2} + \dfrac{1}{3} z^{-2}$	$k_3 = a_3 = \dfrac{1}{3}$	$B_3(z) = z^{-3} A_3(z^{-1}) = \dfrac{1}{3} + \dfrac{5}{8} z^{-1} + \dfrac{13}{24} z^{-2} + z^{-2}$
$A_2(z) = \dfrac{A_3(z) - k_3 z^{-1} B_3(z)}{1 - k_3^2} = 1 + \dfrac{3}{8} z^{-1} + \dfrac{1}{2} z^{-2}$	$k_2 = a_2 = \dfrac{1}{2}$	$B_2(z) = z^{-2} A_2(z^{-1}) = \dfrac{1}{2} + \dfrac{3}{8} z^{-1} + z^{-1}$
$A_1(z) = \dfrac{A_2(z) + k_3 z^{-1} B_2(z)}{1 - k_2^2} = 1 + \dfrac{1}{4} z^{-1}$	$k_1 = a_1 = \dfrac{1}{4}$	$B_1(z) =$ not required

These results are summarized in Figure 11.9, which compares the two solutions (direct and lattice), namely,

$$h_{\text{direct\_FIR}}[k] = \{1, 13/24, 5/8, 1/3\},$$

$$h_{\text{lattice\_FIR}}[k] = \{1, 1/4, 1/2, 1/3\},$$

which produce first few sample values:

Architecture	$y[0]$	$y[1]$
Direct	1	$13/24$
Lattice	1	$1/4 + (1/4)(1/2) + (1/2)(1/3) = 13/24$.

The lattice filter, shown in Figure 11.9, can be expressed as $A(z)$ (Eq. 11.1).

Consider the third-order lattice FIR having PARCOR coefficient $k[i]$ = {0.5, 0.25, 0.125} as shown in Figure 11.10. The transfer function from input to $A_i(z)$ and $B_i(z)$ can be computed using Equations 11.3 and 11.4, producing

$$A_1(z) = 1 + (1/2)z^{-1},$$
$$B_1(z) = (1/2) + 1z^{-1} \text{(reverse)},$$
$$A_2(z) = 1 + (1/2 + 1/2 \times 1/4)z^{-1} + (1/4)z^{-2} = 1 + (5/8)z^{-1} + (1/4)z^{-2},$$
$$B_2(z) = 1/4 + (5/8)z^{-1} + z^{-2} \text{(reverse)},$$
$$A_3(z) = 1 + (1/2 + 1/2 \times 1/4 + 1/4 \times 1/8)z^{-1} +$$
$$(1 \times 1/4 + 1/2 \times 1/4 \times 1/8 + 1/2 \times 1/8)z^{-2} + (1/8)z^{-3}$$
$$= 1 + (21/32)z^{-1} + (21/64)z^{-2} + (1/8)z^{-3} = H(z),$$
$$B_3(z) = 1/8 + (21/64)z^{-1} + (21/32)z^{-2} + z^{-3} \text{(reverse)},$$

where $B_i(z) = z^{-i}A_i(z^{-1})$. The lattice filter's transfer function is

$$A_3(z) = H(z) = 1 + 21/32z^{-1} + 21/64z^{-2} + 1/8z^{-3}$$

or $h[k]$ = { 1, 21/32, 21/64, 1/8}. In the reverse direction, $B_3(z) = 1/8 + (21/64)z^{-1} + (21/32)z^{-2} + z^{-3}$. The magnitude frequency responses of $A(z)$ and $B(z)$ are shown in Figure 11.11, and can be seen to be identical. The difference in the filter responses is found in the phase domain due to their zero locations. The zeros of the $H(z)$ are also computed to be located at $z = -0.5 + 0j$, $-0.078 \pm 0.494j$, and those of the reverse order filter are $z = -2 + 0j$, $-0.312 \pm 1.975j$. They are seen to be reciprocal reflections of each other (i.e., $z = re^{j\phi}$, then $z_R = (1/r)e^{-j\phi}$).

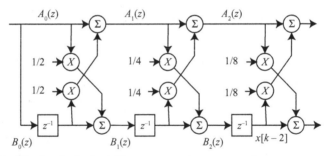

Figure 11.10   Third-order lattice FIR.

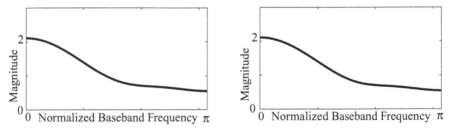

Figure 11.11   Magnitude frequency response of $H(z)$ (left) and the reverse order filter $H_R(z)$ (right).

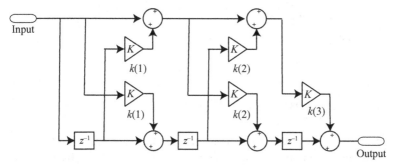

Figure 11.12   Filter Design Toolkit rendering of a fourth-order (third-order MATLAB) lattice FIR architecture (MATLAB style).

Example: MATLAB Lattice FIR

Refer again to a previous lattice example having a transfer function was $H(z) = A_3(z) = 1 + 21/32z^{-1} + 21/64z^{-2} + 1/8z^{-3}$, having PARCOR coefficient $k[i] = \{1/2, 1/4, 1/8\}$ (see Fig. 11.10). The filter's time- and frequency-domain behavior can be examined using MATLAB. The tf2latc command converts a transfer function into a lattice filter, and latc2tf reverses the process. The function latcfilt convolves an input signal with a lattice filter.

Mathworks provides some support in the creation of a lattice FIR architecture using the Filter Design Toolbox dfilt option. In particular, dfilt.latticemamax converts a transfer function into a lattice form FIR. In addition, dfilt.latticemamin is used to convert a transfer function into a minimum phase lattice FIR. The MATLAB generated outcome is displayed in Figure 11.12 and produced using the following MATLAB code:

```
h=[1 21/32 21/64 1/8]; k=tf2latc(h) % transfer function
to lattice conversion {k = 0.5000, 0.2500, 0.1250}

hh=latc2tf(k)   % lattice to transfer function conversion
{hh = 1.0000, 0.6563, 0.3281, 0.1250}

[F,G]=latcfilt(k,[1,zeros(1,1000)]); % convolve input x
with lattice filter producing a forward response F (i.e.,
A(z)) and backwards response G (i.e., B(z)).

[Magf,w]=freqz(F,[1]); [Magg,w]=freqz(G,[1]); % plot
abs(Magf) and abs(Magg)

Hlat=dfilt.latticemamax(k); % Lattice FIR

Hlat

     FilterStructure: 'Lattice Moving-Average (MA) For
Maximum Phase'

         Arithmetic: 'double'
```

```
          Lattice: [0.25 0.50 0.3333333333]

   PersistentMemory: false

realizemdl(Hlat);
```

DISTRIBUTED ARITHMETIC (DA)

Once designed, a filter's coefficients h_i can be assumed to be fixed or constant. As a result, the linear convolution sum-of-product terms are of the form $h_i \times x[i - k]$ that technically define a scaling operation and not multiplication*: As a result, the linear convolution sum consists of a collection of accumulated scaled terms. Whereas multiplication requires the use of a general-purpose multiplier, scaling can be implemented with far lower complexity (e.g., 2's complement [2C] data shift). One of the most attractive alternatives is to replace a traditional multiplier with high bandwidth semiconductor lookup table (LUT) that contains all possible precomputed product (scaled) outcomes. A popular LUT-based technology is called distributed arithmetic (DA). A DA filter assumes that data are coded as an M-bit 2C fractional data word, specifically,

$$x[k] = -x[k:0] + \sum_{i=1}^{M-1} x[k:i]\, 2^{-i}; |x[k]| \le 1. \tag{11.5}$$

Here, $x[k:i]$ denotes the ith bit of sample $x[k]$. Substituting Equation 11.5 into the linear convolution sum (Eq. 6.1), one obtains

$$y[k] = \sum_{r=0}^{N-1} h_r \left(-x[k-r:0] + \sum_{i=1}^{M-1} x[k-r:i]\, 2^{-i} \right)$$
$$= -\sum_{r=0}^{N-1} h_r x[k-r:0] + \sum_{r=0}^{N-1}\sum_{i=1}^{M-1} h_r x[k-r:i]\, 2^{-i}. \tag{11.6}$$

Upon reversing the order of the double summation,* the following results:

$$y[k] = -\sum_{r=0}^{N-1} h_r x[k-r:0] + \sum_{i=1}^{M-1} 2^{-i} \sum_{r=0}^{N-1} h_r x[k-r:i]. \tag{11.7}$$

Suppose that a 2^N-word LUT $\theta[x[k]:i]$ is programmed to contain all possible 2^N values of the inner sum terms in Equation 11.7, namely,

$$\theta[\underline{x}[k]:i] = \sum_{r=0}^{N-1} h_r x[k-r:i]; x[s:i] \in [0,1]. \tag{11.8}$$

---

*Scaling refers to the multiplicative combination of a variable and a constant. Multiplication is defined to be a multiplicative combination of two variables.

*Because of the redistribution of the order of summation, the result is called a DA filter.

All possible 2^N values of $\theta[x[k]:i]$, shown in Equation 11.8, are addressable by an N-bit address vector $x[k:i] = \{x[k:i], x[k-1:i], \ldots, x[k-N-1:i]\}$, where $x[r:i]$ is binary valued (i.e., $\{0, 1\}$). The output word width of the table $\theta[x[k]:i]$ is assumed to be B-bits wide, where B is chosen to meet an output precision requirement. To illustrate, if $N = 8$, assume the data in the ith common bit location is $x[k:i] = \{10110101\}$. Assume also that the filter's impulse response is $h[k] = \{1, 0.9, 0.8, 0.7, 0.6, 0.5, 0.4, 0.3\}$, then $\theta[x[k]:i] = h_0 + h_2 + h_3 + h_5 + h_7 = 1 + 0.8 + 0.7 + 0.5 + 0.3 = 3.3$.

Collectively, the convolution sum, defined by Equation 11.7, becomes

$$y[k] = -\theta[\underline{x}[k]:0] + \sum_{i=1}^{M-1} 2^{-i}\theta[\underline{x}[k]:i] \qquad (11.9)$$

The DA implementation of a convolution sum is seen to consist of a collection of LUT calls, accumulation, and scaling by 2^{-i} as shown in Figure 11.13. Note that the vector of binary values $[x[k]:i]$, for a given i, is presented to the table θ as an address vector beginning with all the least significant bits (LSBs) of $x[k]$, then moving on to the most significant. The first $(M-1)$ binary-valued vectors correspond to positive bits. It is important to note that the bits used to form the vector $[x[k]:i+1]$ have a weight of two relative to the bit values found in $[x[k]:i]$. This explains the scaling factor 2^{-i} found in Equation 11.9, which is implemented as a shift/adder as shown in Figure 11.13. The last address vector is taken from a common sign-bit (negation) location. This requires that the last LUT output θ be subtracted from the accumulator.

In some instances, the maximum real-time bandwidth of a DA FIR can be significantly higher than that of a filter implemented using general-purpose multipliers. Suppose $M = 16$ bits and the filter order is $N = 12$. If the traditional FIR is

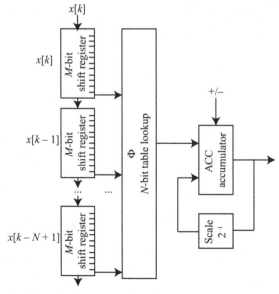

Figure 11.13   Basic Nth-order FIR DA architecture.

implemented using a MAC with a 100-ns cycle time, then the minimum filter cycle time is $T_{\text{FIR\_cycle}} = N \times T_{\text{MAC\_cycle}}$. This results in a 1200-ns cycle time or a real-time rate of 833 kHz. Compared with a DA filter based on a $2^{12} \times 16$ bit memory table having a cycle time of 10 ns, the minimum filter cycle time is $T_{\text{DA\_cycle}} = M \times T_{\text{Memory\_cycle}}$. This translates into a filter cycle time of 160 ns, or a real-time rate of 6.250 MHz, a 750% improvement! This is done without any appreciable increase in hardware complexity and is, in many cases, actually less complex than those designs based on a DSP μp. Because of speed and complexity arguments, DA is often the filter technology of choice for FPGA implementations.

Example: Distributed FIR

Consider a simple fourth-order FIR having a transfer function given by $H(z) = 1.0 - 0.9z^{-1} + 0.64z^{-2} - 0.575z^{-3}$. The worst case unit-bound input is $x[k] = \{1, -1, 1, -1\}$, which would produce a worst case output of $\sum |h_i| = 3.115 < 2^2$. This requires that at least 2 bits be reserved for accumulator "headroom" in order to prohibit run-time register overflow. The largest possible individual table lookup value is $\theta[x[k]:i]_{\max} = 1.64 < 2^1$, which means that the LUT data format must contain at least one integer bit. Assume, for illustrative purposes, that the memory table contents are coded as 8-bit 2C words consisting of a sign bit, 1 integer bit (previously discussed), and 6 fractional bits, to form words having an [8:6] format. The $2^4 = 16$ possible table contents of a $2^4 \times 8$ LUT are precomputed, converted to a 2C word having 6 bits of fractional precision (e.g., $[x[k]:i] = [0,0,1,1] \to <0.64 - 0.575 = 0.065>_6 = 0.0468$ without rounding), and saved (see Table 11.1).

TABLE 11.1. DA Table Contents (2C Truncated Word with 6 Fractional Bits of Precision)

$x[k0:i]$	$x[k1:i]$	$x[k2:i]$	x[k3:i]	Table
0	0	0	0	0
0	0	0	1	−0.5781
0	0	1	0	0.6092
0	0	1	1	0.0468
0	1	0	0	−0.9062
0	1	0	1	−1.4843
0	1	1	0	−0.2656
0	1	1	1	−0.8437
1	0	0	0	0.9843
1	0	0	1	0.4062
1	0	1	0	1.6093
1	0	1	1	1.0468
1	1	0	0	0.0781
1	1	0	1	0.4843
1	1	1	0	0.7187
1	1	1	1	0.1406

TABLE 11.2.   DA Execution Table

i	LUT Address Vector	Table $\langle\theta\rangle_6$	ACC	ACC = ACC/2 ± θ
3	$[x[k]:0\ \mathrm{LSB}] \rightarrow [0001]$	−0.5781	0	$0 + (−0.5781) = −0.5781$
2	$[x[k]:1] \rightarrow [0001]$	−0.5781	−0.5781	$−0.2890 + (−0.5781) = −0.8671$
1	$[x[k]:2] \rightarrow [0001]$	−0.5781	−0.8671	$−0.4335 + (−0.5781) = −1.0116$
0	$[x[k]:3] \rightarrow [0100]$	−0.9062	−1.0116	$−0.5058 − (−0.9062) = 0.40036$

ACC, accumulator.

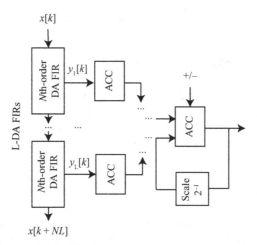

Figure 11.14   High-order DA architecture.

Assume that at sample time $k = 3$, the inputs are four 4-bit input sample values having 2C codes $x[3] = 0 \leftrightarrow [0\Delta000] = 0$, $x[2] = −1 \leftrightarrow [1\Delta000] = −1$, $x[1] = 0 \leftrightarrow [0\Delta000] = 0$, and $x[0] = 1 \leftrightarrow [0\Delta111] = 7/8$ where Δ denotes the binary point location. The real output $y[3]$ is mathematically given by the linear convolution sum $y[3] = h_0x[3] + h_1x[2] + h_2x[1] + h_3x[0] = h_1x[2] + h_3x[0] = 0.3969$ (without rounding). To produce this result using a DA filter, the sequence of operations shown in Table 11.2 is executed.

At the conclusion of the distributed filter cycle, the accumulator holds $y[3] = 0.40039$, which is close to the desired result. The error is 0.003515, an error that is generally smaller than is obtained using a general-purpose arithmetic logic unit (ALU) operating with 6 bits of fractional precision.

Higher-order designs are required when the filter's order exceeds the size of the address space of a single LUT. Higher-order DA filters can be constructed from low-order distributed filters using the tree architecture shown in Figure 11.14. In such cases, the FIR order N is spread across L tables having an address space of n-bits each. In particular $L = \lceil N/n \rceil$, where $\lceil \circ \rceil$ denotes the ceiling function.

TABLE 11.3. 2C to CSD Conversion
(C_i = CSD Value)

c_i	x_{i+1}	x_i	c_{i+1}	C_i
0	0	0	0	0
0	0	1	0	1
0	1	0	0	0
0	1	1	1	−1
1	0	0	0	1
1	0	1	1	0
1	1	0	1	−1
1	1	1	1	0

CANONIC SIGNED DIGIT (CSD)

The CSD system is a ternary-valued (i.e., $\{1, 0, -1\}$) numbering scheme. The CSD was employed in early vacuum tube digital computers and by a first-generation DSP chip (i.e., Intel 2920) in an attempt to accelerate multiplication. A coding comparison of a 2C and CSD word is shown in Table 11.3 and Equation 11.10:

$$x_{(2C)} = -x_n 2^{n-1} + \sum_{i=0}^{n-2} x_i 2^i; \; x_i = \{1, 0\} \leftrightarrow \sum_{i=0}^{n-1} C_i 2^i; \; C_i = \{1, 0, -1\} = x_{(CSD)}. \quad (11.10)$$

The CSD encoding scheme, shown in Table 11.3, is defined in terms the 2C digits x_i, where c_i is a 2C carry-in, c_{i+1} is a 2C carry-out digit, and C_i is the ith CSD ternary-valued digit. The CSD achieves its goal of accelerating multiplication by creating data words that are dense in zeros compared with traditional binary coding schemes (e.g., 2C). The probability of a 2C digit being 0 is 50% but is 67% for the CSD case. This high density in 0 digits translates into an increased "no-op" count and accelerated multiplication. To illustrate, the 4-bit binary representation of the number 15 is given by $15_{10} \leftrightarrow 1111_2$ and requires three shift-adds to complete a sequential multiplication. The CSD representation for 15 is given by $15_{10} = 16_{10} - 1_{10} \leftrightarrow$ CSD and can be implemented using one addition and a hard-wired shift register.

Example: CSD

Code the 3-bit number $x = -2_{10} = [110]_{2C}$ as a CSD number and compute $y = 10x$. From Table 11.3, it follows that

$$c_0 = 0, x_1 = 1, x_0 = 0, c_1 = 0, C_0 = 0$$
$$c_1 = 0, x_2 = 1, x_1 = 1, c_2 = 1, C_1 = -1$$
$$c_2 = 1, x_3 = 1, x_2 = 0, c_3 = N/A, C_0 = 0$$

where $x = C_0 + (2C_1) + (4C_2) = -2$. Therefore, $y = 10x = 10 \times C_0 + 10 \times (2C_1) + 10 \times (4C_2) = -20$.

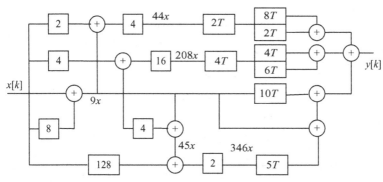

Figure 11.15   RAG implementation of an FIR.

Reduced Adder Graphs (RAG)

The RAG method is also based on the theory of the ternary-valued numbers. The RAG paradigm attempts to extract a power-of-2 factorization of an integer having a minimum number of terms. The cost of a RAG multiplier is measured in terms of the number of adders needed to complete a design. For example, the 2C cost of multiplying by $45_{10} \leftrightarrow 101101_2$ is 3, but using RAG factoring, the cost of a multiply by $45 = 5 \times 9 = (4 + 1) \times (8 + 1)$ is 2. To illustrate the value of a RAG design, consider the half-band filter having an impulse response $h[k] = \{9, 0, -44, 0, 208, 346, 208, 0, -44, 0, 9\}$. To reduce these numbers over common factors, emphasizing radix-2-like decompositions, the data can be represented as follows:

$$346 = 2 \times 173 = 2 \times (128 + 45) = 2 \times (128 + (9 \times 5)) = 2 \times (128 + ((8+1) \times (4+1))),$$
$$208 = 16 \times 13 = 16 \times (4 + 9) = 16 \times (4 + (8+1)),$$
$$44 = 4 \times 11 = 4 \times (2 + 9) = 4 \times (2 + (8+1)),$$
$$9 = 8 + 1,$$

which is graphically interpreted in Figure 11.15.

Example: RAG and CSD Implementation

A compensator is to be designed to account for the roll-off attributed to an in-line filter as shown in Figure 11.16. The compensator is to have approximately a reciprocal magnitude frequency response over the in-line filter's passband. The 15th-order FIR compensator's impulse response is given (as integers) by $h[k] = \{-1, 4, -16, 32, -64, 136, -352, 1312, -352, 136, -64, 32, -16, 4, -1\}$. The CSD coefficient code is shown below. It can be seen that the low-complexity RAG FIR improves the overall system magnitude frequency response. It can be seen that both filters have a low complexity. The CSD-enabled FIR filter cycle would require 2×5 one-digit coefficients plus 2×1 two-digit coefficients plus 1 three-digit coefficient. The RAG implementation is based on radix-2 coefficients 1, 2, 4, 8, 16, 32, and 64, along with hard-wired 2C shifts.

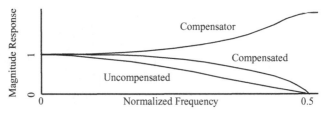

Figure 11.16   Compensating FIR filter's magnitude frequency response.

Coefficient	CSD	RAG
1312	1024 + 256 + 32	32 × (32 + 8 + 1)
352	−230 64 ??	−32 × (8 + 2 + 1)
136	128 + 8	8 × (8 + 1)
64	64	64
−64	−64	−64
−16	−16	−16
4	4	4
−1	−1	1

FIR FINITE WORD LENGTH EFFECTS

It may be recalled that the worst case FIR gain can be easily computed. It shall be assumed that the input has been properly scaled so that no run-time overflow errors occur. Errors that occur within such a system are then attributed to finite word length effects in the form

- coefficient round-off errors, or
- arithmetic rounding errors.

Coefficient round-off errors, as the name implies, correspond to errors associated with converting real numbers with a digital approximation having F-bits of fractional precision. The difference between the coefficients is called the coefficient round-off error. It is assumed that the rounding of a set of real filter coefficient h_i can be modeled as $h_i^R = h_i + \Delta_i$, where $|\Delta_i| \le 2^{-(F+1)}$, where $|\Delta_i| \le 2^{-(F+1)}$ (i.e., ±LSB/2). The size of the error is seen to be a direct function of F, the number of fractional bits used in the finite word length representation of h_i. Based on this model, the linear convolution outcome of an ideal (real valued) and fixed-point system can be compared. The linear convolution of an FIR having an impulse response h_i with an input signal $x[k]$ produces an outcome $y[k]$. Convolving the input with a set of fixed-point FIR filter coefficients results in an output response,

$$y^R[k] = \sum_{m=0}^{N-1} \left(h_m + \Delta_m \right) x[k-m] = y[k] + \sum_{m=0}^{N-1} \Delta_m x[k-m].$$ (11.11)

The error, due to coefficient rounding is

$$e[k] = (y'[k] - y[k]) = \sum_{m=0}^{N-1} \Delta_m x[k-m].$$ (11.12)

The error budget is seen to be scaled by the individual values of $x[k]$. It is often assumed that the input time series has a mean value of zero (i.e., $E(x[k]) = 0$). If the input is assumed to be an impulse, that is, $x[k] = \delta[k]$, then Equation 11.12 simplifies to

$$e[k] = (y'[k] - y[k]) = \sum_{m=0}^{N-1} \Delta_m \delta[k-m] = \Delta_k,$$ (11.13)

where Δ_k comes from a uniformly distributed random population defined over $[-2^{-F/2}, 2^{-F/2})$. In this case, the error is seen to have a mean value of zero ($E(e[k]) = 0$) and a variance $\sigma^2 = 2^{-2F}/12$ (i.e., $Q^2/12$, Q being the quantization step size). This is of little practical value, however, since the input signal is rarely just an impulse. If the input is a random or arbitrary signal, then the error can be modeled as

$$\sigma_e^2 = N\sigma_\Delta^2 \sigma_x^2 = \frac{N2^{-2F}\sigma_x^2}{12},$$ (11.14)

where σ_x^2 is the signal variance (power). Therefore, the coefficient round-off error variance of an Nth-order FIR is essentially the round-off error power associated with each rounding (i.e., $Q^2/12$), scaled by the signal power and filter order. For example, the predicted round-off error variance of a 64th-order FIR is expected to be twice that of a 32nd-order filter. A standard means of analyzing these data is to interpret the error in bits ($\log_2(\sigma_e)$). Therefore, going from a 32nd-order to a 64th-order FIR costs only 1 bit of precision due to coefficient rounding.

The coefficient round-off error associated with DA or CSD filters is application dependent. DA filters are less susceptible to coefficient rounding because the sum of products, defined by Equation 11.8 (i.e., $\theta[x[k]:i]$), is computed using floating-point precision, and then is rounded to a fixed-point number. The coefficient round-off errors introduced by CSD encoding are dependent on the original coefficient distribution. Some coefficients lend themselves to exact CSD coding, others resist.

The coefficient round-off error can also be examined in the context of a signal-to-noise ratio, or SNR, where the SNR (in decibels) is given as

$$\text{SNR} = 10 \times \log_{10}\left(\frac{\sigma_h^2}{\sigma_e^2}\right) = 10 \times \log_{10}\left(\frac{\sigma_h^2}{2^{-2F}/12}\right),$$ (11.15)

where σ_h^2 denotes the variance (power) of the FIR's impulse response $h[k]$, and σ_e^2 is the variance (power) in the error process $e[k] = h[k] - h_F[k]$, where $h_F[k]$ is $h[k]$ rounded to F-fractional bits of precision. Based on this model, the improvement in statistical SNR performance of adding one additional fractional bit to the coefficients would be

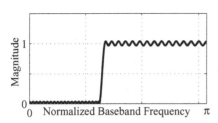 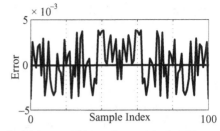

Figure 11.17 Magnitude frequency response of a high-pass 101st-order equiripple FIR (left) and coefficient round-off errors for 7 fractional bits ($\Delta = 2^{-7}$) (right).

$$10\log_{10}\left(\frac{\sigma_x^2/\left(2^{-2F}/12\right)}{\sigma_x^2/\left(2^{-2(F+1)}/12\right)}\right) = 10\log_{10}\left(\frac{12}{4}\right) = 4.77 \text{ dB/bit.} \qquad (11.16)$$

Example: Coefficient Round-Off Error

An analysis of a 101-order equiripple high-pass FIR, having a passband cut-off frequency of $0.2125f_s$ and a stopband cut-off frequency of $0.2f_s$, is shown in Figure 11.17. The analysis includes determining the worst case gain (2.54), maximum tap-weight coefficient (0.587), and the effects of coefficient rounding in an SNR context. Assume, for analysis purposes that the filter is to be implemented using an 8-bit data format. If an [8:7] fixed-point data format is chosen, then all the tap-weight coefficients can be successfully coded ($|h[i]| < 1$). Assume further that the input is also bounded in magnitude to be less than unity (i.e., $|x[k]| < 1$). The full-precision tap-weight multiplier outputs are 16-bit words having a sign bit and $F = 15$ bits of fractional precision (i.e., [16:15]). However, the worst case gain calculation $G = 2.54 < 2^2$ indicates that two additional bits of accumulator "headroom" would be needed to remove any chance of run-time overflow. This means that the accumulator output format should be [16 + 2:15] = [18:15]. If providing an extended precision accumulator is impractical, then to ensure that run-time overflow does not occur, the input, FIR coefficients, or combination of input and coefficients would need to be scaled by at least $G = 2.54$, with an attendant loss in output precision. In any case, the accumulator's output would normally be returned to a common format (e.g., 8 bits) at the end of the filter cycle. These ideas are explored in the MATLAB code insert, shown below. The results are summarized in Figure 11.17 and the execution of the code is shown below. At 7 fractional bits, the 4.77-dB guideline would have predicted the output SNR is $7 \times 4.77 = 33$ dB, which is close to the value computed (30.04 dB):

```
h=firpm(100,[0 .4 .425 1],[0 0 1 1]);  % equiripple FIR

norm(h,1)  % worst case gain: L1 norm = 2.5437

max(abs(h))  % largest coefficient 0.5875

t=0:100; h8=round(2^7*h)/2^7;  % round to [8:7]

SNR=10*log10(var(h)/var(h-h8));  % SNR = 30.0436
```

One observation that should be kept in mind is that linear phase FIR filter coefficients tend to be defined in terms of small valued numbers found at the beginning and end of the impulse response support (called tapers). In many fixed-point designs ($N < 16$ bits), these coefficients can be easily rounded to zero. Therefore, while increasing the order of an FIR to meet design specification may seem like a good idea when using floating-point design tools (e.g., MATLAB), many of the new tap-weight coefficients can be inadvertently quantized to zero.

Example: Coefficient Round-Off Error

A 21st-order equiripple linear phase low-pass FIR has a direct architecture implementation shown in Figure 11.18. The coefficients of the resulting floating-point FIR is converted into a fixed-point filter using an [8:7] format. The design is equipped with full-precision multipliers and an extended precision (by 2 bits) accumulator. The filter coefficients of the floating- and fixed-point filter are compared. The floating and fixed-point filter performances are compared in Figure 11.19. It can be seen that finite word length effects do have an effect, albeit marginal, on the frequency response of the fixed-point system. The zero distributions were also seen to be subject to minor changes due to finite word length effects.

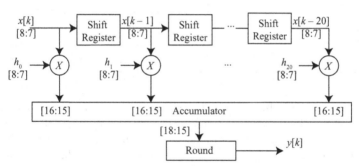

Figure 11.18 Architecture of a 21st-order FIR showing coefficients rounded to 7 fractional bits ($\Delta = 2^{-7}$), full precision multipliers, and extended precision accumulator.

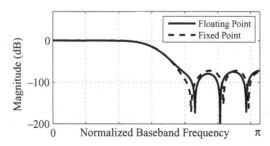

Figure 11.19 Comparison of floating- and fixed-point magnitude frequency responses.

ARITHMETIC ERRORS

Arithmetic errors can occur when the partial products $x[k - m] \times h_m$ being produced by a convolution operation are performed using finite precision arithmetic. Mathematically quantifying this process is, in general, a challenging problem. The errors associated with an ideal multiplier, operating on fixed-point operands X and Y, are given by

$$\text{var}(XY) = (E(X))^2 \text{var}(Y) + (E(Y))^2 \text{var}(X) + \text{var}(X)\text{var}(Y). \qquad (11.17)$$

To appreciate this process, consider forming a digital product of two variables X and Y, coded as 2C fixed-point variables X_r and Y_r. Suppose two 4-bit 2's complement numbers, having a [4:3] format, are to be multiplied. For $X_{10} = 6/8$ ($6/8_{2C} = 0110$) and $Y_{10} = -5/8$ ($-5/8_{2C} = 1011$), resulting in the product $-30/64_{10}$. The computational process is illustrated below:

{extended sign bit}	0	0110 0000	{form first partial product–rightmost bit = LSB}
{extended sign bit}	0	0011 0000	{shift right preserving extended sign bit}
{extended sign bit}	0	1001 0000	{add second partial product}
{extended sign bit}	0	0100 1000	{shift right, preserving extended sign bit}
{extended sign bit}	0	1001 0000	{add third partial product = 0}
{extended sign bit}	0	0010 0100	{shift right, preserving extended sign bit}
{extended sign bit}	1	1100 0100	{subtract fourth partial product = 0}
{extended sign bit}	1	1110 0010	{shift right, preserving extended sign bit}
		1110 0010	{discard extended sign bit, final value = −30}

The outcome is seen to be an 8-bit 2C word having an [8:7] format. The value of the LSB also has changed from 2^{-3} to 2^{-7}. The data can, at this point, be interpreted as a double-precision data word or rounded to a single-precision entity. What needs to be statistically measured is the quality of the outcome. This question can be explored using computer simulation, which can be used to multiply two long uniformly distributed random processes represented as real and [16:15] fixed-point data words. The experimental results are presented below:

```
x1=2*(rand(1,1000)-0.5);x2=2*(rand(1,1000)-0.5) % random
inputs x1 and x2

x1r=round((2^15)*x1)/2^15; x2r=round((2^15)*x2)/2^15; %
quantize to [16:15]

log10(sqrt(var(x1-x1r)))/log10(2)    % quantization error
in bits -16.7950

log10(sqrt(var(x2-x2r)))/log10(2)    % quantization error
in bits -16.8099

y=x1.*x2; yr=x1r.*x2r; % sum of products
```

```
log10(sqrt(var(y-yr)))/log10(2)    % quantization error in
bits -17.0831
```

First, quantizing the input data to 15 fractional bits produces a result with a predictable quantization error. For $\Delta = 2^{-15}$, the predicted error is $-15 - 1.79 = -16.79$ fractional bits, which is close to the computed quantization errors of -16.795 and -16.81 for inputs $x_1[k]$ and $x_2[k]$, respectively. The quantized and real arrays are multiplied and their difference is compared. The statistical difference of -17.08 bits is found to be less than the individual input quantization errors. While the error is better than that of a [16:15] word, it is far from achieving the LSB value of the double-precision [32:31] word. To examine the nature of these errors requires that the basic MAC structures, as shown in Figure 11.20, be understood. The MAC #1 architecture performs error-free real multiplication and accumulation. MAC #2 performs a full-precision digital multiply and then rounds the outcome to a single-precision outcome that is then accumulated using single-precision arithmetic. One of the issues with the MAC #2 architecture is that the accumulator needs to be checked for overflow at the end of each accumulation. This slows the accumulation process and adds data-dependent stalls to any pipeline. To overcome this obstacle, the adder shown in the MAC #2 is redesigned as MAC #3. This design can be implemented using an extended precision accumulator having a word width of $(N + R)$ bits, where $R = \lceil \log_2(G) \rceil$) where G is the filter's worst case gain. For a typical FIR, R will have a value of but a few bits. For example, if $G = 2.54 < 2^2$, then two additional bits of accumulator "headroom" would be needed to eliminate any chance of run-time overflow. This means that the accumulator output format should be $[N + 2:F]$. The price paid would be an increase in complexity. What is gained is a filter that is protected from run-time register overflow and the elimination of run-time overflow detection operations. Finally, if the MAC #3 filter performs one rounding operation per filter cycle (reduction of a $2N$ or $2N + R$-bit product to an N-bit outcome), the resulting output quantization error estimate is given by $\sigma^2 = NQ^2/12$, $Q = 2^{-F}$, which is normally several bits less than a round-after-multiply architecture.

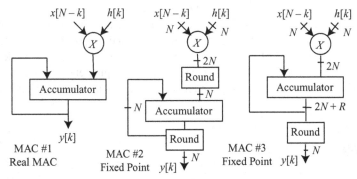

Figure 11.20   Basic MAC structures. MAC #1 (left), MAC #2 (middle), and MAC #3 (right).

Example: MAC errors

The three MAC architectures shown in Figure 11.20 are experimentally explored using a 21st-order linear phase low-pass FIR and a unit-bound random noise input. The outputs, for both real and [16:15] fixed-point instances are compared. The analysis begins with the calculation of worst case gain of $G = 1.771 < 2^1$, which indicates that the extended precision accumulator requires 1 bit of headroom. The maximum FIR coefficient is found to be 0.4487, which means all the coefficients can be coded as a [16:15] 2C word. In some instances, design engineers advocate using a [16:16] format, which effectively means that the FIR coefficients are scaled (upward) by a factor of 2 before coding. This, in concept, would add one additional bit of precision to the coded coefficients. However, it should be noted that the worst case gain is now increased to $G = 3.54$, which will require an additional 2 bits (instead of 1) of headroom to protect against run-time overflow. The result is the increased 1-bit precision that is taken back by the need for 1 additional integer (head room) bit. The experimental results are summarized below, where the error is compared with the ideal (floating-point) filter response:

Case	$x[k]$	$h[k]$	$y[k]$	$\log_2(\sigma)$	Comment
y_0	Real	Real	Real	$F = -\infty$	Ideal filter response
y_1	[16:15]	[16:15]	Real	$F = -15.5$	Precision for an ideal filter processing quantized input data. Slightly degraded over a basic [16:15] error of -16.79 bits.
y_r	Real	Real	[16:15]	$F = -16.8$	Real input and fixed-point FIR. Error agrees with theoretical prediction of -16.79 bits.
y_n	[16:15]	[16:15]	[16:15]	$F = -15.3$	Precision for a fixed-point filter processing quantized input data. Slightly degraded over a basic [16:15] error of -16.79 bits.

The arithmetic errors associated with DA or CSD filters represent a set of special cases. A DA FIR is known to enjoy a precision advantage over direct or transpose FIR implementations. An Nth-order DA FIR is assumed to have the statistical error budget given by $\sigma^2 = Q^2/9$ versus $\sigma^2 = NQ^2/12$ for direct architecture models. The errors introduced by a computer-aided design (CAD) implementation are essentially due entirely to coefficient rounding, and arithmetic errors are zero if extended precision adders are used.

SCALING

The worst case gain of an FIR filter gain is denoted G_{max}. Intuitively, it would make sense for the worst case gain to have a power-of-two value, say 2^n. This can be achieved by scaling the FIR by a factor k where $k = 2^n/G_{max}$. By multiplying each filter coefficient h_i by k, the new FIR would have a maximum gain of $G'm_{ax} = 2^n$ as a result. This action would be performed off-line and not affect the filters' run-time performance. It should be appreciated, however, that the scaled filter's magnitude frequency response will be increased by a factor k. Scaling can take place in a downward direction as well. Logically, a scale factor $k = 2^{n-1}/G_{max}$ can also be considered. The scale factor k can be used to adjust the gains downward to the nearest power-of-two value. While the absolute magnitude frequency response will be changed, the relative gain will be left unaffected.

MULTIPLE MAC ARCHITECTURES

As technology provides more powerful and compact arithmetic devices, multiple MACs architectures are becoming a viable option. FIRs typically require multiple MAC calls to be performed per filter cycle. This provides motivation for implementing multiple MAC FIR architectures. Suppose M MACs are available for use in implementing an Nth-order FIR convolution sum, where $x[k]$ is a bounded input time series. The filter tap-weight coefficients, denoted h_k, are assumed to be real. Upon presenting an input to the FIR, an output time series $y[k]$ will result having the form

$$y[0] = h[0]x[0]$$
$$y[1] = h[1]x[0] + h[0]x[1]$$
$$\cdots = \cdots$$
$$y[N-1] = h[N-1]x[0] + h[N-2]x[1] + \cdots + h[0]x[N-1].$$

For sample indices $k \geq N - 1$, N physical multiplications must be performed per filter cycle. These multiples can be performed sequentially or concurrently. If M MACs are available for concurrent use, and N is divisible by M, then a $K = N/M$-fold speed up can be realized. Suppose, however, that $N = KM + K_0$, $K_0 \in [1, N-1]$, then $K + 1$ MAC cycles would be required where the first K cycles would use all M MACs concurrently and the last cycle would use only K_0 of the available MACs. The efficiency of this action can be mathematically represented by Δ:

$$\Delta = \frac{\text{actual speed-up}}{\text{ideal speed-up}} = \frac{L/M}{\lceil L/M \rceil}, \qquad (11.18)$$

where $\lceil \circ \rceil$ again denotes the ceiling function. The value of Δ is interpreted in Figure 11.21 for other cases. Notice that the efficiency improves as L increases, which simply reflects the fact that the overhead is reduced when a large number of MAC need to be performed.

The multiple MAC filters shown in Figure 11.22 are called horizontal (direct) and vertical (transpose) architectures since the multiple MACs are spread horizontally and vertically across the convolution space. It is assumed that the coefficients are loaded into registers and are spatially attached to each MAC. Each one of the

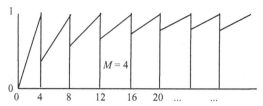

Figure 11.21   Speed-up potential of a four-multiplier FIR as a function of filter order.

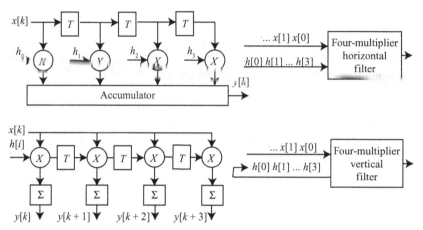

Figure 11.22   Example of a four-multiplier horizontally (direct) architected FIR.

TABLE 11.4.   **Four-Multiplier Vertically (Transpose) Architected FIR Execution**

Clock	Cell 0	Cell 1	Cell 2	Cell 3	Sum/Clear
0	$h_3x[0]$	0	0	0	0
1	$h_2x[1]$	$h_3x[1]$	0	0	0
2	$h_1x[2]$	$h_2x[2]$	$h_3x[2]$	0	0
3	$h_0x[3]$	$h_1x[3]$	$h_2x[3]$	$h_3x[3]$	Cell 0 = $y[3]$
4	$h_3x[4]$	$h_0x[4]$	$h_1x[4]$	$h_2x[4]$	Cell 1 = $y[4]$
5	$h_2x[5]$	$h_3x[5]$	$h_0x[5]$	$h_1x[5]$	Cell 2 = $y[5]$
6	$h_1x[6]$	$h_2x[5]$	$h_3x[5]$	$h_0x[5]$	Cell 3 = $y[6]$
7	$h_0x[7]$	$h_3x[6]$	$h_2x[6]$	$h_3x[6]$	Cell 0 = $y[7]$

M MACs could be dedicated to the computation of $y[k]$ at one sample instance. This form is called a vertical architecture that is graphically interpreted in Figure 11.22. Again the efficiency of the multi-MAC architecture is a function of the relationship between the filter length N and the number of MACs M. The execution of the first several clock instances of a four-multiplier vertically architected FIR is shown in Table 11.4.

CHAPTER 12

CLASSIC FILTER DESIGN

INTRODUCTION

There are instances when a low filter order and/or a filter having precise frequency selectivity are the overriding design considerations. In such cases, a finite impulse response filter (FIR) may not be a viable option. The antithesis of an FIR is the infinite impulse response (IIR) filter. IIRs can often provide a low-complexity, frequency-selective solution. An IIR system is modeled in terms of a transfer function having the form

$$H(z) = \sum_{k=0}^{\infty} h_k z^{-k} = \frac{\displaystyle\sum_{k=0}^{M} b_k z^{-k}}{\displaystyle\sum_{k=0}^{N} a_k z^{-k}} = K z^{N-M} \frac{\displaystyle\prod_{i=0}^{M-1}(z - z_i)}{\displaystyle\prod_{i=0}^{N-1}(z - p_i)} = \frac{N(z)}{D(z)}. \qquad (12.1)$$

It can be seen that an IIR has both numerator $N(z)$ and denominator $D(z)$ terms. This is in sharp contrast with an FIR that possesses only numerator terms (i.e., no feedback). The presence of the denominator $D(z)$ indicates that an IIR contains feedback giving rise to a filter having an impulse response that can persist forever.

The baseband frequency response of an IIR filter is determined by evaluating the transfer function $H(z)$ in the z-domain along the periphery of the unit circle at points $z = e^{j\varpi}$, $-\pi \le \varpi \le \pi$. The frequency response is formally given by

$$H(e^{j\varpi}) = \frac{\displaystyle\sum_{i=0}^{M} b_i e^{-j\varpi i}}{\displaystyle\sum_{i=0}^{N} a_i e^{-j\varpi i}} = K e^{j\varpi(N-M)} \frac{\displaystyle\prod_{i=0}^{M-1}(e^{j\varpi} - z_i)}{\displaystyle\prod_{i=0}^{N-1}(e^{j\varpi} - p_i)}. \qquad (12.2)$$

It is claimed that an IIR can meet very demanding frequency response requirements and excel in those cases where the frequency response must be highly selective. However, IIRs have many shortcomings as well. Due to the presence of feedback, IIRs are far more susceptible to encounter instability and arithmetic overflow problems during run time when compared with an FIR. Improperly managing these

Digital Filters: Principles and Applications with MATLAB, First Edition. Fred J. Taylor.
© 2012 by the Institute of Electrical and Electronics Engineers, Inc.
Published 2012 by John Wiley & Sons, Inc.

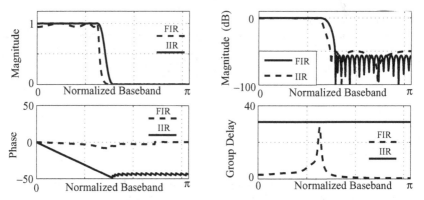

Figure 12.1 Comparison of a 63rd-order linear phase FIR and sixth-order IIR showing magnitude frequency responses (top left), logarithmic (decibels) magnitude frequency responses (top right), phase responses (bottom left), and group delay responses (bottom right).

conditions can lead to a disastrous outcome. Other IIR issues include nonlinear phase performance (variable group delay) and coefficient round-off error sensitivity.

Example: Comparison of IIRs and FIRs

The magnitude frequency and phase responses of a 63rd-order linear phase FIR filter and a sixth-order low-pass IIR filter are compared in Figure 12.1. It can be seen that both filters have similar magnitude frequency responses. The IIR's phase response of the IIR is seen to be nonlinear resulting in a frequency-dependent propagation delay (i.e., group delay). Because the IIR is 1/10th the order of the FIR, it can potentially run 10 times faster.

CLASSIC ANALOG FILTERS

IIR filters can be defined in terms of their transfer function $H(z)$ in the z-domain. It may be recalled that a z-transform is based on an established relationship between the s- and z-domains. It is therefore reasonable to explore analog filters where solutions can be interpreted in the context of discrete-time or digital filters. Classic analog filters have been studied and chronicled for nearly a century. Their origins can be traced to the early days of radio, where frequency-selective filters were required as a radio infrastructure technology. Classic analog filters were defined in terms of an ideal low-pass, high-pass, band-pass, band-stop, and all-pass frequency response models. The body of knowledge associated with the design of classic analog filters is enormous. Historically, classic analog filters are generally based on Bessel, Butterworth, Chebyshev, and Cauer (elliptic) filter models. Compared with Butterworth, Chebyshev, and Cauer filters, Bessel filters have a relatively flat group delay (linear phase like). However, since true linear phase behavior can be obtained with a simple FIR, digital Bessel filters are rarely implemented today. The other filter types, however, are commonly found in today's digital technology. The design of a digital Butterworth, Chebyshev and Cauer (elliptic) filter begins with the specifications of a classic analog filter. This information can be translated into filter design parameters. Today, like yesterday, the analog IIR filter design strategy is a highly structured procedure. For a classic design, the first step is to define an analog low-pass filter having a normalized critical passband frequency of $\Omega = 1$ rad/s that reflects many of the desired filter's attributes. Such a filter is called an analog prototype filter and has a transfer function $H_p(s)$. The next step accounts for the fact that the desired filter can be something other than low-pass (e.g., band-pass) with a $\Omega = 1$ rad/s passband. The prototype filter is mapped to its final analog form $(H_p(s) \rightarrow H(s))$ using a mapping rule called a frequency–frequency transform (see Table 12.1 and Fig. 12.2).

Example: Frequency–Frequency Mapping

A third-order −3-dB passband Butterworth analog low-pass prototype model, having a 1 rad/s critical frequency, is given by

TABLE 12.1.   Frequency–Frequency Transforms

Nth-Order Prototype	Frequency–Frequency Transform	Final Order
Low-pass to low-pass	$j\Omega \leftarrow j\Omega/\Omega_p$	N
Low-pass to high-pass	$j\Omega \leftarrow \Omega_p/j\Omega$	N
Low-pass to band-pass	$j\Omega \leftarrow (j\Omega^2 + \Omega_H\Omega_L)/(j\Omega(\Omega_H - \Omega_L))$	$2N$
Low-pass to band-stop	$j\Omega \leftarrow (j\Omega(\Omega_H - \Omega_L))/(j\Omega^2 + \Omega_H\Omega_L)$	$2N$

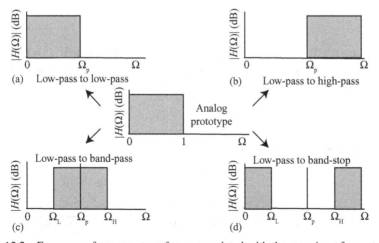

Figure 12.2   Frequency–frequency transforms associated with the mapping of a prototype low-pass into (a) low-pass, (b) high-pass, (c) band-pass, and (d) band-stop filters.

$$H_\mathrm{p}(s) = \frac{1}{s^3 + 2s^2 + 2s + 1}.$$

The filter's DC gain is $|H_\mathrm{p}(j0)| = 1$ and the gain at the critical frequency $s = j\Omega = j1$ is $|H_\mathrm{p}(j1)| = 0.707$, corresponding to the −3-dB point. In a practical setting, the actual −3-dB frequency would normally be something other than 1 rad/s, say 1 kHz. Using a low-pass-to-low-pass frequency–frequency transformation (Table 12.1), for $\Omega_\mathrm{p} = 2\pi \times 1000 = 6280$ rad/s, results in an analog filter model given by

$$H(s) = H_\mathrm{p}(s)\big|_{s \to s/\Omega_\mathrm{p}} = \frac{1}{s^3 + 2s^2 + 2s + 1}\bigg|_{s \to s/\Omega_\mathrm{p}}$$

$$= \frac{1}{4.03 \times 10^{-12} s^3 + 5.07 \times 10^{-8} s^2 + 3.18 \times 10^{-4} s + 1}$$

Evaluating $H(s)$ at $s = j\Omega_\mathrm{p} = j6280$ rad/s (i.e., 1 kHz) verifies that the filter's gain is $|H(j6280)| = 0.707$ or −3 dB at the low-pass cutoff frequency as required.

PROTOTYPE ANALOG FILTERS

The key to designing a classic digital filter (Butterworth, Chebyshev, elliptic) is deriving the transfer function of the analog prototype filter. Much is known about this process, which has been reduced to a set of algebraic and computer instructions. The design procedure for a classic analog prototype is presented below for the purpose of completeness. In all cases, the low-pass analog prototypes filters are synthesized to meet magnitude frequency response specifications as shown in Figure 12.3. The analog low-pass prototype is shown to have a magnitude squared frequency response, measured at $\Omega = 1$ rad/s, to be

$$|H(s)|^2_{s=j1} = \frac{1}{1+\varepsilon^2}. \tag{12.3}$$

If $\varepsilon^2 = 1$, or $\varepsilon = 1$, the prototype is said to be a −3-dB filter. If $\varepsilon^2 = 0.2589$, or $\varepsilon = 0.508$, the prototype is said to be a −1-dB filter. Other design parameters are defined in terms of the data shown in Figure 12.3. The passband and stopband gains

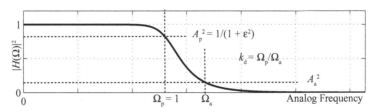

Figure 12.3   Magnitude frequency response squared of a low-pass analog prototype filter having a critical frequency of 1 rad/s.

are specified in terms of the parameters ε and A_a, respectively. The depth of the filter's skirt can be measured in terms of the transition gain ratio given by $\eta = \varepsilon / \sqrt{A_a^2 - 1}$ or the alternative reciprocal gain ratio $d = 1/\eta$. The frequency transition ratio, denoted k_d, is a measure of the width of the transition bandwidth. The value of k_d is case dependent and given by

$$\text{low-pass: } k_d = \frac{\Omega_p}{\Omega_a}, \tag{12.4}$$

$$\text{high-pass: } k_d = \frac{\Omega_a}{\Omega_p}, \tag{12.5}$$

$$\text{band-pass: } k_d = \begin{cases} k_1 & \text{if } \Omega_u^2 \geq \Omega_v^2 \\ k_2 & \text{if } \Omega_u^2 < \Omega_v^2 \end{cases}, \tag{12.6}$$

$$\text{band-stop: } k_d = \begin{cases} 1/k_1 & \text{if } \Omega_u^2 \geq \Omega_v^2 \\ 1/k_2 & \text{if } \Omega_u^2 < \Omega_v^2 \end{cases}. \tag{12.7}$$

From this list of parameters, the order and transfer function of classical analog prototype filters can be determined for Butterworth, Chebyshev, and Cauer/elliptic filter models.

BUTTERWORTH PROTOTYPE FILTER

An Nth-order Butterworth prototype filter has a magnitude squared frequency response given by

$$|H(\Omega)|^2 = \frac{1}{1 + \varepsilon^2 (\Omega/\Omega_p)^{2N}}. \tag{12.8}$$

The Butterworth IIR is sometimes referred to as a maximally flat filter because the first $2N - 1$ derivatives of $|H(\Omega)|^2 = 0$ at $\Omega = 0$. The prototype Butterworth IIR has a critical frequency $\Omega_p = 1$ rad/s and is used as the foundation for designing Butterworth digital filters having user-specified critical frequencies. The order of a Butterworth IIR can be defined in terms of the design parameters, including $|H(\Omega_a)|^2 = A_a^2$ (see Fig. 12.3). The filter order is estimated to be

$$N = \frac{1}{2} \frac{\log[(A_a - 1)/\varepsilon^2]}{\log(\Omega_s/\Omega_p)} = \frac{\log(d)}{\log\left(\frac{1}{k_d}\right)} = \frac{\log\left(\frac{1}{\eta}\right)}{\log\left(\frac{1}{k_d}\right)}. \tag{12.9}$$

The $2N$ order transfer function squared $|H(s)|^2$ can be factored into $|H(s)|^2 = H(s)H(-s)$, where $H(s)$ is called the realizable filter containing all the stable poles (i.e., left-hand plane poles) of $|H(s)|^2$, and $H(-s)$ is the unstable filter. The s-domain poles of $|H(s)|^2$ are located on a circle of radius $r = 1/\varepsilon^{1/N}$ and are separated by π/N radians. As

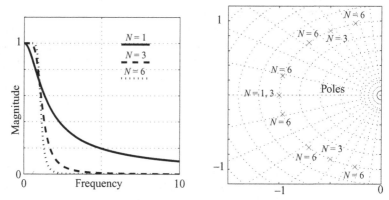

Figure 12.4 Magnitude frequency response of an order 1, 3, and 6 order analog −3-dB Butterworth analog prototype filters (left), and pole locations of an order 1, 3, and 6 order analog Butterworth prototype (right).

a result, the poles of a classic Butterworth are located at $s_k = re^{-j(2k-1)\pi/2N}$, $k \in [0, 2N - 1]$, resulting in N stable pole locations and N unstable pole locations. To illustrate, three −3-dB ($\varepsilon = 1$) Butterworth analog low-pass prototype filters are designed of order $N = 1$, 3, and 6. All three Butterworth prototypes have a critical passband frequency of $\Omega = 1.0$ rad/s. The production of these filters can be facilitated using the MATLAB `butter` and `buttord` (order) commands and are shown in Figure 12.4. The MATLAB-derived −3-dB Butterworth prototype filters are

$$H_1(s) = \frac{1}{s+1},$$

$$H_3(s) = \frac{1}{s^3 + 2s^2 + 2s + 1},$$

$$H_6(s) = \frac{1}{s^6 + 3.86s^5 + 7.46s^4 + 9.13s^3 + 7.46s^2 + 3.86s + 1}.$$

The pole locations of a prototype Butterworth filter are distributed along the periphery of a unit circle in the s-domain. The pole of the first-order Butterworth prototype is located at $s = -1$, the third-order Butterworth prototype has poles at -1, and $-\cos(30°) \pm j \sin(30°)$, and the sixth-order Butterworth prototype poles are located at $s_k = e^{j[\pi/2 + \pi/12 + (2k\pi/12)]}$ for $k \in [1, 6]$.

Example: Butterworth Prototype

Design of a low-pass Butterworth filter with a 35-Hz ($\Omega_p = 2\pi \times 35$) −1-dB ($\varepsilon = 0.508$) passband and $A_a = 10^3$ (−60 dB) stopband attenuation starting at 100 Hz ($\Omega_a = 2\pi \times 100$) (after Fig. 12.3). The remaining design parameters are given by $\eta = \varepsilon/(A_a^2 - 1)^{1/2} = 0.508 \times 10^{-3}$ and $k_d = 35/100 = 0.35$. The filter order can therefore be estimated to be

$$N = \frac{\log\left(\dfrac{1}{0.508 \times 10^{-3}}\right)}{\log\left(\dfrac{1}{0.35}\right)} = 7.22 < 8$$

or order 8. The filter order N could have been machine-calculated using the MATLAB `buttord` command. The resulting eighth-order prototype Butterworth filter is given by

$$H_p(s) = \frac{1}{s^8 + 5.125s^7 + 13.137s^6 + 21.846s^5 + 25.688s^4 + 21.846s^3 + 13.137s^2 + 5.125s + 1}.$$

Next, the prototype filter $H_p(s)$ is mapped into a final analog filter using a low-pass-to-low-pass frequency–frequency transform of the form $s \rightarrow s/\Omega_p = s/(2\pi 35)$ (see Table 12.1). The resulting transfer function for the final analog Butterworth low-pass filter is given by

$$H(s) = \frac{1.1 \times 10^{19}}{s^8 + 1.23 \times 10^3 s^7 + 7.53 \times 10^5 s^6 + 3.00 \times 10^8 s^5 + 8.42 \times 10^{10} s^4}$$
$$+ 1.72 \times 10^{13} s^3 + 2.47 \times 10^{15} s^2 + 2.31 \times 10^{17} s + 1.1 \times 10^{17}$$

The result of an eighth-order Butterworth filter has a −1-dB passband of 35 Hz and a stopband beginning at 100 Hz as shown in Figure 12.5.

CHEBYSHEV PROTOTYPE FILTER

The magnitude squared frequency response of an Nth-order prototype low-pass Chebyshev I filter is given by

$$|H(s)|^2 = \frac{1}{1 + \varepsilon^2 C_N^2(\Omega/\Omega_c)}, \tag{12.10}$$

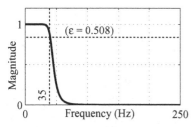 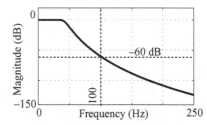

Figure 12.5 Magnitude frequency response of eighth-order Butterworth filter ($\Omega_p = 220$ rad/s [35 Hz] and $\Omega_a = 628$ rad/s [100 Hz]) (left) and the log magnitude frequency response (right).

where $C_N(s)$ is an Nth-order Chebyshev polynomial given by

$$C_N(s) = \begin{cases} \cos(N\cos^{-1}(s)), |s| \leq 1 \\ \cosh(N\cosh^{-1}(s)), |s| > 1 \end{cases} \tag{12.11}$$

and satisfying the recursive relationship

$$C_{N+1}(s) = 2sC_N(s) - C_{N-1}(s). \tag{12.12}$$

The first few Chebyshev polynomials are $C_0(s) = 1$, $C_1(s) = s$, $C_2(s) = 2s^2 - 1$, $C_3(s) = 4s^3 - s$, and so forth. The order of a Chebyshev I filter, in terms of filter specifications, is estimated to be

$$N = \frac{\log\left[\dfrac{1+\sqrt{1-\eta^2}}{\eta}\right]}{\log\left[\dfrac{1}{k_d} + \sqrt{\left(\dfrac{1}{k_d}\right)^2 - 1}\right]}. \tag{12.13}$$

The $2N$ poles of $|H(s)|^2 = |H(s)H(-s)|$ lie on an ellipse whose geometry is determined by ε and N. Once again, N of the poles $|H(s)|^2$ belongs to the stable left-hand plane and is assigned to $H(s)$. In particular, the left-hand plane poles are located at $s_k = -\sin(x[k])\sinh(y[k]) + j\cos(x[k])\cosh(y[k])$, where $x[k] = (2k - 1)\pi/2N$, and $y[k] = \pm\sinh^{-1}(1/\varepsilon)/N$. The other N unstable poles belong to $H(-s)$.

A variation on the Chebyshev I model is called the Chebyshev II filter. The Chebyshev II magnitude squared frequency response is given by

$$|H(j\omega)|^2 = \frac{1}{1+\left[\varepsilon^2\dfrac{C_N^2(\Omega_a/\Omega)}{C_N^2(\Omega/\Omega_a)}\right]}. \tag{12.14}$$

The order estimation formula (Equation 12.14) is the same for both the Chebyshev I and II models. A typical magnitude frequency response for a Chebyshev I or II filter is displayed in Figure 12.6. The differential features between the Chebyshev I and II filter strategies can be deduced from the displayed responses. The Chebyshev

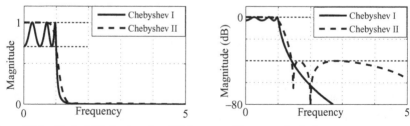

Figure 12.6 Typical Chebyshev I and II low-pass filter magnitude frequency response (left) and in decibel units (right).

I is seen to exhibit equiripple behavior in the passband and have a smooth transition into a relatively flat stopband. The Chebyshev II filter is seen to have an equiripple stopband and smooth transition into a relatively flat passband.

Example: Chebyshev Low-Pass Filter

Design a Chebyshev I and II that meets a set of specifications of a prototype filter. The gain specifications include a stopband bound of $A = 10^3$ (−60 dB) and a passband gain limited by $\varepsilon = 0.508$ (−1 dB). The passband width is 35 Hz and the stopband begins at 100 Hz. The remaining design parameters are given by $\eta = \varepsilon/(A^2 - 1)^{1/2} = 0.508 \times 10^{-3}$ and $k_d = 35/100 = 0.35$. The filter order is estimated to be

$$N = \frac{\log\left[\left(\dfrac{1 + \sqrt{1 - 0.258 \times 10^{-6}}}{0.508 \times 10^{-3}}\right)\right]}{\log\left[2.857 + \sqrt{7.6163}\right]} = 4.83 < 5.$$

These specifications resulted in an eighth-order Butterworth IIR. The fifth-order −1-dB prototype Chebyshev I and II filters (denoted CI and CII) are given by

$$H_{CI}(s) = \frac{0.122}{s^5 + 0.94s^4 + 1.69s^3 + 0.97s^2 + 0.58s + 0.122},$$

$$H_{CII}(s) = \frac{0.0142 \times (s^4 + 32.65s2 + 213.25)}{s^5 + 4.00s^4 + 7.89s^3 + 9.73s^2 + 7.52s + 3.05}.$$

After applying a low-pass-to-low-pass frequency–frequency transform, the final version of the Chebyshev I and II filters become

$$H_{CI}(s) = \frac{6.32 \times 10^{10}}{s^5 + 2.06 \times 10^2 s^4 + 8.17 \times 10^4 s^3 + 1.04 \times 10^7 s^2 + 1.35 \times 10^9 s + 6.32 \times 10^{10}},$$

$$H_{CII}(s) = \frac{3.14(s^4 + 1.58 \times 10^6 s^2 + 5.00 \times 10^{11})}{s^5 + 8.74 \times 10^2 s^4 + 3.82 \times 10^5 s^3 + 1.04 \times 10^8 s^2 + 1.76 \times 10^{10} s + 1.56 \times 10^{12}}.$$

The frequency responses of the analog Chebyshev I and II filters are shown in Figure 12.7.

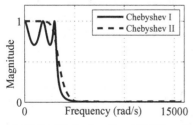
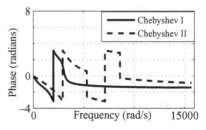

Figure 12.7   Chebyshev I and II filters showing magnitude frequency (left) and phase responses (right).

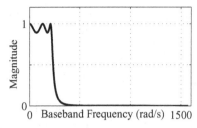 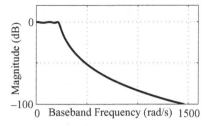

Figure 12.8 Magnitude frequency response of a fifth-order Chebyshev I low-pass filter for $\Omega_p = 220$ rad/s (35 Hz) and $\Omega_a = 628$ rad/s (100 Hz) (left) and log-magnitude frequency response (right).

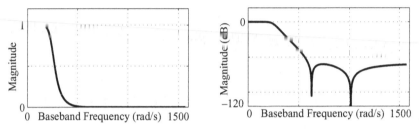

Figure 12.9 Magnitude frequency response of a fifth-order Chebyshev II filter for $\Omega_p = 220$ rad/s (35 Hz) and $\Omega_a = 628$ rad/s (100 Hz) (left) and log-magnitude frequency response (right).

The design algorithms and design process can be reduced to computer programs. Specifically, a fifth-order Chebyshev I and II design can be mechanized using MATLAB. The design process computes the required filter order using the `cheb1ord` and `cheb2ord` commands. A Chebyshev I or II IIR can then be synthesized using MATLAB's `cheby1` and `cheby2` commands. The designed fifth-order Chebyshev I filter's transfer function can be produced using the command `tf(B,A)`, with the following result:

$$H(s) = \frac{6.31 \times 10^{10}}{s^5 + 2.06 \times 10^2 s^4 + 8.16 \times 10^4 s^3 + 1.03 \times 10^7 s^2 + 1.35 \times 10^9 s + 6.31 \times 10^{10}},$$

which is in agreement with the previously derived solution. The response of the resulting Chebyshev I filter is shown in Figure 12.8.

The fifth-order Chebyshev II filter is similarly designed and displayed in Figure 12.9. Both filters essentially have the same gross magnitude frequency response. At the microscopic level, the MATLAB Chebyshev I exhibits passband ripple while the Chebyshev II filter has ripple in the stopband.

ELLIPTIC (CAUER) PROTOTYPE FILTER

An Nth-order prototype elliptic filter is expressed in terms of a Jacobian elliptic integral equation of the form

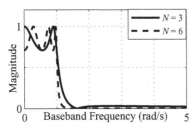 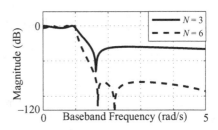

Figure 12.10   Typical elliptic filter magnitude frequency response (left) and in decibels (right).

$$E(x) = \int_0^{\pi/2} \frac{d\theta}{\sqrt{1 - x_2 \sin^2(\theta)}}. \tag{12.15}$$

The order of an elliptic filter is estimated to be0

$$N \geq \frac{\log(16D)}{\log\left(\dfrac{1}{q}\right)}, \tag{12.16}$$

where the other parameters are

$$k' = \sqrt{(1 - k_d^2)}, q_0 = \frac{1 - \sqrt{k'}}{2(1 + \sqrt{k'})}, q = q_0 + 2q_0^5 + 15q_0^9 + 15_q^{13}, D = d^2; d = 1/\eta. \tag{12.17}$$

A typical magnitude frequency response of an elliptic low-pass filter is presented in Figure 12.10. It can be seen that an elliptic filter exhibits equiripple behavior in both the pass- and stopbands.

Example: Elliptic Prototype

Design an elliptic filter's stopband gain that is bounded by $A = 10^3$ (−60 dB) beginning at 100 Hz and the passband gain is defined by $\varepsilon = 0.508$ (−1 dB) out to 35 Hz. The remaining design parameters are given by $\eta = \varepsilon/(A^2 - 1)^{1/2} = 0.508 \times 10^{-3}$ and $k_d = 35/100 = 0.35$. The filter order is given by

$$k' = \sqrt{(1 - 0.35^2)} = 0.9367,$$

$$q_0 = \frac{1}{2}\frac{(1 - \sqrt{k'})}{(1 + \sqrt{k'})} = 0.0081667,$$

$$q \approx q_0 = 0.0081667,$$

$$D = d^2 = 1968^2; d = 1/\eta.$$

$$N \approx \frac{\log(16D)}{\log\left(\frac{1}{q}\right)} = 3.73 < 4.$$

These specifications require an eighth-order Butterworth IIR of fifth-order Chebyshev I and II IIRs. Unlike a Butterworth or Chebyshev filter, the system's poles are not easily expressed in terms of a formula. The filter coefficients can be machine-computed using the MATLAB `ellip` function, and the order determined by using the `ellipord` command. The fourth-order Elliptic filter's transfer function can be produced using MATLAB's `tf(B,A)` command. Using manual techniques, the analog prototype fourth-order −1-dB elliptic prototype filter is computed to be

$$H_p(s) = \frac{0.000461(s^4 + 22.502s^2 + 65.237)}{s^4 + 0.553s^3 + 0.498s^2 + 0.149s + 0.033}.$$

After applying the low-pass-to-low-pass frequency–frequency transform, the final version of the elliptic filter emerges and is given by

$$H(s) = \frac{0.000461(s^4 + 3.229 \times 10^6 s^2 + 1.342 \times 10^{12})}{s^4 + 2.096 \times 10^2 s^3 + 7.148 \times 10^4 s^2 + 8.148 \times 10^6 s + 6.961 \times 10^8}.$$

The zeros are found at $s = \pm j1654.7$ and $\pm j700.3$ while the poles are located at $s = -75.76 \pm j93.21$ and $-29.065 \pm j217.7$. The elliptic filter's magnitude frequency response is shown in Figure 12.11.

PROTOTYPE TO FILTER CONVERSION

The process of designing a classic analog Butterworth, Chebyshev I–II, and elliptic filters first involves designing an analog prototype filter $H_p(s)$. This filter is then mapped into its final form (i.e., $H(s)$) using frequency–frequency transforms. The resulting filter $H(s)$ meets or exceeds the posted analog frequency domain specifications in terms of gains, and gain deviations at the critical frequencies. This process is motivated by the following example.

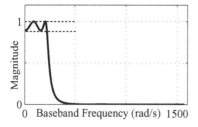
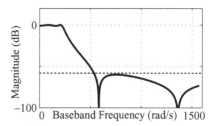

Figure 12.11   Magnitude frequency response of a fourth-order elliptic analog filter for $\Omega_p = 220$ rad/s (35 Hz) and $\Omega_a = 628$ rad/s (100 Hz) (left) and log-magnitude frequency response (right).

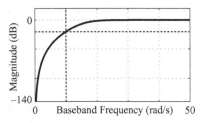 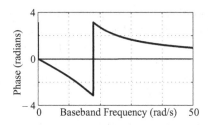

Figure 12.12 Analog high-pass Butterworth filter. Magnitude frequency response (left) and phase response (right).

Example Design

Design an analog high-pass Butterworth IIR filter having a −20 dB or better stopband ranging over $\Omega \in [0, 10]$ rad/s, and a passband with an allowable 0.1-dB deviation beginning at $\Omega = 40$ rad/s. The manual filter design process begins with producing an Nth-order analog low-pass prototype model having a critical frequency of 1 rad/s (MATLAB `buttord` command). This filter is then mapped to a final high-pass form $H(s)$ using a low-pass high-pass frequency–frequency transform (MATLAB `lp2hp` command). The filter can also be machine-computed using MATLAB. The filter order can be determined to be $N = 4$ and computed using MATLAB's `buttord` command. The Butterworth high-pass filter can be computed using MATLAB's `butter` command. The resulting low-pass prototype filter's transfer function can be produced using MATLAB's `tf(B,A)` command to produce

$$H_{HP}(s) = \frac{s^4}{s^4 + 46s^3 + 1077s^2 + 14639s + 99499}.$$

The outcome is graphically shown in Figure 12.12.

OTHER IIR FILTER FORMS

The purpose of Butterworth, Chebyshev I and II, and elliptic filter design protocols is to construct a good facsimile to a piecewise constant ideal filter magnitude frequency response model. Analog filter structures, other than the classical Butterworth, Chebyshev I and II, and elliptic filter models are required at times. Filters with an arbitrary magnitude frequency response can be defined by the synthesis techniques based on estimation tools, such as autoregressive (AR) or autoregressive moving average (ARMA) models. In all cases, the objective is to synthesize an Nth-order analog filter having a transfer function of the form

$$H(s) = \frac{\sum_{m=0}^{M} b_m s^m}{\sum_{m=0}^{N} a_m s^m} \tag{12.18}$$

whose frequency response agrees with the observed or measured output of the system being modeled in some acceptable manner. Experienced analog filter designers also work directly in the s-plane. By adjusting the location of the poles and zeros, relative to the $j\Omega$-axis, the frequency response of an analog filter can be shaped until it satisfies the designer's requirements. Again, the result is a transfer function $H(s)$.

Regardless of the design strategy used to synthesize an analog filter $H(s)$, it needs to be converted into the z-domain in order to realize a digital filter. There are basically two strategies that commonly map an s-domain transfer function $H(s)$ into a z-domain transfer function $H(z)$, resulting in a digital filter. These techniques are called the impulse invariant and bilinear z-transform methods. However, some digital filter design paradigms are not based on analog models; instead, they convert digital filter specifications into a final form using what are called direct synthesis methods. Some of the more common direct methods are summarized below.

PRONY'S (PADÉ) METHOD

The majority of IIR starts are defined in terms of a set of frequency domain specifications. There are, however, times when the filter is defined in terms of an arbitrary time-domain set of specifications. In such a case, Prony's method can be used to convert a set of time-domain specifications into an IIR design having the form

$$H(z) = \frac{\sum_{i=0}^{M} b_i z^{-i}}{\sum_{i=0}^{N} b_i z^{-i}} = \frac{B(z)}{A(z)} = \sum_{i=0}^{\infty} h[i] z^{-i}. \tag{12.19}$$

An L sample impulse response can be used to generate an IIR $H(z)$, where $L \geq N + M + 1$, using MATLAB's Prony's tool.

Example: Prony's Method

Synthesize the transfer function $H(z)$ from knowledge of the first 26 samples of the IIR's impulse response. Suppose that the original IIR is a fourth-order low-pass Butterworth filter having a normalized passband edge of 0.2π. The synthesis process is illustrated below as

```
[b,a]=butter(4,0.2); % Design a 4th order Butterworth
IIR

h=filter(b,a,[1 zeros(1,25)]); % Create L=26-point impulse
response

[B,A]=prony(h,4,4); % Synthesize an IIR for N=M=4,
L>M+N+1
```

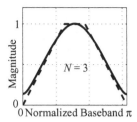 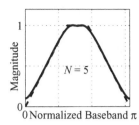 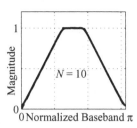

Figure 12.13   Yule–Walker designs for filter orders 3, 5, and 10.

The measured initial 26 impulse response samples of the fourth-order Butter-worth filter is the time series $h[k] = \{0.0048, 0.0307, 0.0906, 0.1679, \ldots, -0.0010, 0.0004\}$, which upon being processed by Prony, for $N = M = 4$, returns a filter model:

$$H(s) = \frac{0.0048s^4 + 0.019s^3 + 0.0289s^2 + 0.019s + 0.0048}{s^4 - 2.369s^3 + 2.314s^2 - 1.054s + 0.187},$$

which corresponds to the coefficients of the original fourth-order Butterworth filter.

YULE–WALKER

Another direct design method is based on the Yule–Walker equations. The MATLAB program yulewalk designs an IIR digital filter, which has a magnitude frequency response that best approximates that of a measured or specified magnitude frequency response. The desired filter order is user selected.

Example: Yule–Walker

Design a 3rd-, 5th-, and 10th-order IIR that have a magnitude frequency response that best approximates

$$\left| H(e^{jn\Omega}) \right| = \begin{cases} n/4; \, n = 0,1,2,3 \\ 1; \, n = 4,5,6 \\ (n-10)/4 : n = 7,8,9,10 \end{cases}.$$

The frequency responses of the MATLAB-realized 3rd-, 5th-, and10th-order IIR filter, based on the Yule–Walker equations, are shown in Figure 12.13. They are seen to provide a reasonable facsimile of the desired response up to the limits of their order.

CHAPTER 13

IIR DESIGN

INTRODUCTION

The process of designing a classic Butterworth, Chebyshev, or elliptic filter begins with the production of a continuous-time transfer function $H(s)$. Once created, $H(s)$ needs to be converted into a discrete-time filter (a.k.a., digital filter $H(z)$). As it turns out, there are a number of ways in which this mapping can be accomplished. There are, however, two universally accepted techniques, called the impulse invariant transform and the bilinear z-transform method that are in common use. These methods generally produce decidedly different outcomes.

IMPULSE INVARIANCE

One of the obvious techniques that can be used to convert a classical analog filter $H(s)$ into a discrete-time filter $H(z)$ is the standard z-transform. The standard z-transform is also referred to as an impulse invariant transform. An impulse invariant transform preserves the connection between a system's continuous-time and discrete-time impulse responses at the sample instances. Suppose the impulse response of a continuous-time filter having a transfer function is $h_a(t)$. Then, upon sampling the continuous-time impulse, a discrete-time impulse response $h_d[k]$ results. The impulse invariant property of the standard z-transform insures that $h_d[k] = h_a(t = kT_s)$ for all k. Furthermore, the z-transform of the discrete-time impulse response $h_d[k]$ can be expressed in terms of the Laplace transform of the continuous-time impulse response $h_a(t)$. Specifically,

$$H(z)\big|_{z=e^{sT_s}} = \frac{1}{T_s} \sum_{n=-\infty}^{\infty} H_a\left(s + j\frac{2\pi}{T_s}n\right). \tag{13.1}$$

Evaluating Equation 13.1 along the periphery of the unit circle in the z-domain produces the discrete-time filter frequency response. Specifically,

$$H\left(e^{j\varpi}\right) = \frac{1}{T_s} \sum_{n=-\infty}^{\infty} H_a\left(j\frac{\varpi}{T_s} + j\frac{2\pi}{T_s}n\right), \tag{13.2}$$

Digital Filters: Principles and Applications with MATLAB, First Edition. Fred J. Taylor.
© 2012 by the Institute of Electrical and Electronics Engineers, Inc.
Published 2012 by John Wiley & Sons, Inc.

for $|\varpi| \le \pi/T_s$. The term $2\pi n/T_s$ appearing in Equation 13.2 reinforces the fact that signal energy at analog frequencies above the Nyquist frequency, namely, $|\varpi| = \pi/T_s$, is mapped into the discrete-time filter's baseband. This effect is historically called aliasing. If the discrete-time system is to be alias free, then $H_a(j\varpi/T_s + 2\pi n/T_s) = 0$ for all $n \ne 0$. This can be approximately accomplished using an anti-aliasing filter. Under this condition,

$$H\left(e^{j\varpi}\right) = \frac{1}{T_s} \sum_{n=-\infty}^{\infty} H_a\left(j\frac{\varpi}{T_s}\right). \tag{13.3}$$

A problem arises when one considers a practical analog filter. Analog filters generally have a finite magnitude response that persists for all frequencies. For illustrative purposes, consider the simple integrator having a transfer function $H(s) = 1/s$ and a frequency response $H(j\Omega) = 1/j\Omega$. For any finite sample rate $f_s = 1/T_s$, the integrator has a nonzero gain having a magnitude $|H(j\Omega)| = 1/|\Omega|$. In the context of Equation 13.2, the frequency response of the discrete-time filter will contain energy passed by an analog filter at frequencies above the Nyquist frequency. Consider the experiment shown in Figure 13.1, which displays the magnitude frequency response of a typical analog low-pass filter with a 1 rad/s passband shown. Suppose the sample rate is $f_s = 5/2\pi$ Sa/s; then, the low-pass filter's gain has a nonzero value for frequencies residing above the Nyquist frequency (2.5 rad/s). As a result, signal components beyond 2.5 rad/s will be aliased back into the discrete-time filter's baseband. This highlights a serious design limitation that can compromise the potential value of an impulse invariant digital filter. Specifically, they can have difficulty in meeting specific frequency-domain requirements due to excessive aliasing. This caveat does not necessarily diminish the importance of impulse invariant filters in that there are many instances where the digital filter is designed as an analog replacement having known time-domain characteristics. The acceptance criteria, in this case, are developing a solution having the time-domain behavior of the continuous-time system being replaced.

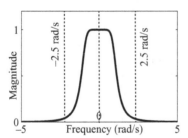 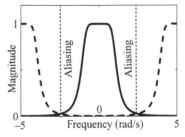

Figure 13.1 Spectrum of a fourth-order low-pass impulse invariant Butterworth filter with a 1 rad/s passband. Shown are the magnitude frequency response of the parent analog filter (left) and the spectrum of the impulse invariant digital filter sampled at $f_s = 5$ Sa/s after Equation 13.2 (right). Note that the spectral overlap introduces aliasing.

IMPULSE INVARIANT DESIGN

An impulse invariant filter is based on an analog parent filter having an assumed transfer function $H_a(s)$ and impulse response $h_a(t)$ given by

$$H_a(s) = \sum_{m=1}^{N} \frac{A_m}{(s - s_m)} \xrightarrow{\text{Inverse Laplace}} h_a(t) = \sum_{m=1}^{N} T_s A_m e^{s_m t} u(t). \qquad (13.4)$$

Upon sampling the continuous-time impulse response, a discrete-time time series results having a standard z-transform transfer function $H_d(z)$ given by

$$h_d[k] = h_d[kT_s] = \sum_{m=1}^{N} A_m e^{s_m kT_s} u(t)$$

$$= \sum_{m=1}^{N} A_m \left(e^{s_m T_s}\right)^k u(t) \xrightarrow{z\text{-transform}} H_d(z) = \sum_{m=1}^{N} \frac{T_s A_m}{\left(1 - e^{s_m T_s} z^{-1}\right)}. \qquad (13.5)$$

From the impulse invariant property of z-transforms, it follows that $h_a(t)|_{t=kT_s} = h_d[k]$ results. Refer to Figure 13.1 and assume that the impulse invariant version of an analog filter is to be used as an analog filter replacement. It is desired that the time-domain behavior of the z-transformed filter closely approximates that of the parent's analog system. In this scenario, it is assumed that the digital device is an analog replacement technology. It has been established, however, that aliasing can compromise the digital solution. To illustrate, consider a simple first-order continuous-time system with a time constant $\tau = 10^{-2}$ and $H(s) = (1/\tau)/(s + [1/\tau])$. When sampled at the high rate of $f_s = 1000$ Sa/s, the filter's magnitude frequency responses are comparable out with the Nyquist frequency (see Fig. 13.2). At the lower sample rate of $f_s = 250$ Sa/s, there is a notable difference (see Fig. 13.2). The reason for this difference is aliasing. Analog or continuous-time filters have stopbands that typically fall off slowly (e.g., −20 dB per decade for a first-order filter). Specifically, the analog filter's gain around the Nyquist frequency is between 0.1 and 0.2, and has nonzero gain beyond that frequency. All the energy that resides beyond the Nyquist frequency will be returned to baseband as an aliased signal. The lower the sample

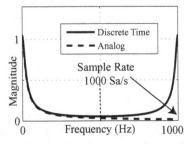
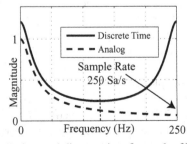

Figure 13.2   Simulated response of a continuous-time and discrete-time first-order filter at sample rates of $f_s = 1000$ Sa/s (left) and $f_s = 250$ Sa/s (right).

rate, the greater the aliasing. When the filter is, however, sufficiently oversampled, such as $f_s = 1000$ Sa/s, then the Nyquist frequency resides deeper in the analog filter's stopband and proportionally less aliasing occurs. The next example explores this observation.

Example: Second-Order Impulse Invariant Filter

A second-order -1-dB Butterworth low-pass discrete-time filter having a 1-kHz passband is designed. The filter's transfer function is given by

$$H(s) = \cfrac{1.965}{\cfrac{s^2}{(2000\pi)^2} + 1.9822\cfrac{s}{2000\pi} + 1.9652} = \frac{7.757 \times 10^7}{s^2 + 1.2454 \times 10^4 s + 7.757 \times 10^7}$$

having poles located at s $= -6.22 \times 10^3 \pm j6.22 \times 10^3$. The system's impulse response has the general form

$$h(t) = e^{-at}\sin(\omega_0 t)u(t) = \left(-0.5je^{(-a+j\omega_0)t} + 0.5je^{(-a-j\omega_0)t}\right).$$

The analog filter can be used to develop a discrete-time impulse invariant filter by applying the standard z-transform and selecting a sample rate. For $f_s = 5$ kHz (five times passband), the discrete-time response and transfer function is

$$h[k] = \alpha^k \sin(\beta k T_s) \xleftarrow{\ z\ } \frac{2\alpha z \sin(\beta T_s)}{T_s\left(z^2 - 2\alpha z \cos(\beta T_s) + \alpha^2\right)},$$

where $T_s = 1/f_s = 2 \times 10^{-4}$, $\alpha = e^{-6.228 \times 10^3 T_s} = 0.2878$, and $\beta = 6.228 \times 10^3$. The transfer function is

$$H(z) = \frac{Kz}{(z - 0.0919)^2 + (0.272686)^2} = \frac{Kz}{z^2 - 0.1960z + 0.0828},$$

where the value of K ($K = 0.679$) is computed using the z-transform rules and K is also chosen ($K = 0.8990$) so that the filter's 0 Hz (DC) gain is unity (i.e., $H(z = 1)$). The magnitude frequency responses are simulated for sample rates of $f_s = 5$ kHz and shown in Figure 13.3. There is a notable difference between the analog and impulse invariant discrete-time filter's magnitude frequency responses due to aliasing.

The process by which an analog filter is mapped into an impulse invariant discrete-time filter can be automated using MATLAB's impinvar command. This process maps an analog filter, denoted $H(s)$, into a discrete-time filter $H(z)$ with a given sample rate.

Consider the previously studied analog filter $H(s) = 7.757 \times 10^7/(s^2 + 1.2454 \times 10^4 s + 7.757 \times 10^7)$ sampled at a 5 kSa/s rate. The continuous-time filter $H(s)$ can be mapped into the impulse invariant filter using the standard z-transform $H(z) = Ts \times 3.3968 \times 10^3 z^1/(z^2 - 0.1968z + 0.0828) = 0.67936z^1/(z^2 - 0.1968z + 0.0828)$ resulting in the outcome shown in Figure 13.3.

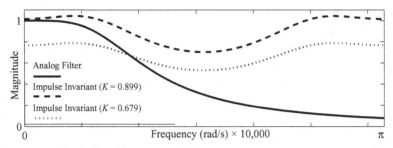

Figure 13.3   Simulated responses of a second-order continuous- and discrete-time impulse invariant Butterworth filter. The numerator gains are $K = 0.8990$ and 0.6793. The energy in the analog filter's frequency response beyond the Nyquist frequency is aliased back into the baseband of the discrete-time filters.

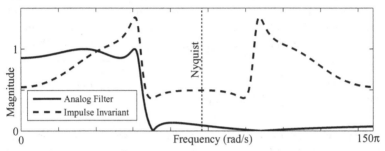

Figure 13.4   Comparison of a fourth-order low-pass elliptic analog and impulse invariant filter's magnitude frequency responses. Data are displayed over a frequency range $\omega = [0, \omega_s]$ (rad/s).

Example: Higher-Order Example

A fourth elliptic low-pass filter is designed with a 25-Hz-wide, 1-dB passband and a 20-dB stopband attenuation. The transfer function $H(s)$

$$H(s) = \frac{0.1s^4 + 1.312 \times 10^{-14} s^3 + 1.338 \times 10^4 s^2 + 7.497 \times 10^{-10} s^1 + 3.2 \times 10^8}{s^4 + 142s^3 + 4.137 \times 10^4 s^2 + 3.367 \times 10^6 s^1 + 3.59 \times 10^8}.$$

For $f_s = 75$ Sa/s, the impulse invariant filter is given by

$$H(z) = \frac{-0.188z^4 + 0.4871z^3 + 0.617z^2 + 0.5476z^1 + 0.0002009}{z^4 + 0.72 + 0.7312z^3 + 0.8305z^2 + 0.02649z^1 + 0.1506}.$$

The impulse invariant filter's impulse response can be produced using MATLAB's function `impz` command. The frequency response difference between the analog parent and the filter's impulse invariant version is shown in Figure 13.4 and is due to aliasing.

Discrete-time mathematics and simulations demonstrate that impulse invariant filters have a difficult time realizing quality magnitude frequency response (see Figs.

13.3 and 13.4). The problem is that practical impulse invariant filters suffer from aliasing. Aliasing occurs because analog filters have finite gain beyond a practical and/or acceptable Nyquist frequency. As a result, the impulse invariant z-transform is generally not recommended for use in implementing infinite impulse responses (IIRs) that must satisfy a set of frequency-domain specifications. However, there are times when the IIR design specifications are given in the time domain, such as rise time and overshoot. In this application domain, impulse invariance is a definite asset. Nevertheless, an alternative to the standard z-transform is needed if frequency selective IIRs are to be a reality. This is the role of the bilinear z-transform.

BILINEAR Z-TRANSFORM

While the standard z transform filter was shown to possess the impulse invariant property, it was found to be problematic when compliance with frequency-domain specifications is a design prerequisite. Problems in meeting frequency-domain goals were attributed to the effects of aliasing. Aliasing effects are anticipated from knowledge that

1. the spectrum of any periodically sampled signal is periodically replicated on frequency centers, which are multiples of the sample frequency f_s (Eq. 13.1), and

2. the gain of an analog filter never actually falls to zero at or beyond a selected meaningful Nyquist frequency.

As a consequence, it should be assumed that the highest frequency component found in the analog filter's response is infinity. Therefore, it would appear that any finite sampling frequency will technically result in aliasing. Aliasing effects can, however, be mitigated using interpolation of some form. A potentially superior interpolation form is a simple integrator, an operator that functions as a low-pass filter. A continuous-time integrator's input–output transfer function is given by $H(s) = 1/s$. A discrete-time model of the integrator is the classic Reimann integrator given by $y[k + 1] = y[k] + (T_s/2) (x[k] + x[k + 1])$. In the z-domain, the Reimann integrator can be expressed as

$$H(z) = \frac{Y(z)}{X(z)} = \frac{T_s}{2} \frac{\left(1 + z^{-1}\right)}{\left(1 - z^{-1}\right)} = \frac{T_s}{2} \frac{(z + 1)}{(z - 1)}. \tag{13.6}$$

Fusing together $H(s)$ with $H(z)$, one obtains

$$s = \frac{2}{T_s} \frac{(z - 1)}{(z + 1)} \text{ or } z = \frac{\left((2/T_s) + s\right)}{\left((2/T_s) - s\right)}. \tag{13.7}$$

Equation 13.7 is called a bilinear z-transform or Tustin's method. One factor that immediately stands out is that the bilinear z-transform is a simple algebraic mapping rule between the continuous-time and discrete-time domains. This is in sharp contrast with a more complicated standard z-transform's exponential mapping (i.e.,

$z = e^{sT_s}$). However, this by itself is insufficient to consider using the bilinear z-transform as a replacement to the venerable standard z-transform. Recall that the problem associated with the standard z-transform was one of aliasing. That is, the frequency components residing in the "tail" of an analog filter, extending beyond the Nyquist frequency, are aliased back into the discrete-time filter's baseband. Overcoming this problem would be sufficient justification for using the bilinear z-transform as a replacement for the standard z-transform. But first, the frequency-domain properties of the linear z-transform (Eq. 13.7) need to be understood.

WARPING

The bilinear z-transform distorts (called warping) the discrete-time filter's frequency axis. To examine the effects of warping, assume that the frequency response of an analog filter is defined by $H_a(j\Omega)$. Let the normalized frequency response of a discrete-time filter be given by $H(e^{j\varpi})$, where $\varpi \in [-\pi, \pi]$. The discrete-time frequency axis can be calibrated, if necessary, with respect to the actual sampling frequency using $\varpi = \omega/T_s$. The problem at hand is to construct a mapping from the s-plane to the z-plane in a manner that eliminates aliasing. To demonstrate, consider evaluating Equation 13.7 along the trajectory $s = j\Omega$ in the s-domain and the arc $z = e^{j\varpi}$ in the z-domain. Direct analysis yields

$$j\Omega = \left(\frac{2}{T_s}\right)\frac{e^{j\varpi}-1}{e^{j\varpi}+1} = \left(\frac{2}{T_s}\right)\frac{j\sin(\varpi/2)}{\cos(\varpi/2)} = \left(\frac{2}{T_s}\right)(j\tan(\varpi/2)). \qquad (13.8)$$

Upon simplification, Equation 13.8 can be expressed as

$$\Omega = \left(\frac{2}{T_s}\right)\tan\left(\frac{\varpi}{2}\right) \text{ or } \varpi = 2\tan^{-1}(\Omega T_s/2). \qquad (13.9)$$

These relationships are interpreted in Figure 13.5.

The mapping rule that takes the normalized baseband frequency ϖ to Ω (Eq. 13.9) is called the prewarping equation. The mapping rule that takes Ω to the normalized baseband frequency ϖ (Eq. 13.9) is called the warping equation. It can be seen that the relationship between the analog- and discrete-frequency axes is nonlinear. As a result, the critical frequencies of an analog filter model will not, in general, align themselves with the critical frequencies of the resulting discrete-time filter. This mismatch can be managed using the prewarping and warping equations. What is of significance is how the bilinear z-transform overcomes the aliasing problem that limited the impulse invariant design method. It can be seen that as the analog frequency increases toward $\Omega \to \infty$ along the continuous-time frequency axis (i.e., $j\Omega$), $\omega \to \pi f_s$ along the discrete-time frequency axis (i.e., $j\omega$). This point corresponds to a normalized Nyquist frequency $\varpi \to \pi$. That is, the entire positive and negative analog frequency axis is now uniquely mapped onto the unit circle. As a result, no aliasing is possible. Because of this property, the bilinear z-transform is

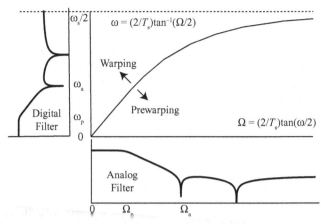

Figure 13.5 Relationship between the continuous and discrete-frequency axes under a bilinear z-transform.

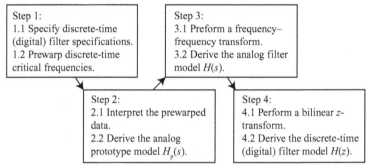

Figure 13.6 Design process for a discrete-time IIR from an analog model using a bilinear z-transform.

well suited to convert classic analog filters into a discrete-time IIR model that preserves the shape of the magnitude frequency response of the parent analog system. The design paradigm is, unfortunately not a straightforward process due to the fact that the mapping rules are nonlinear. The bilinear z-transform procedure is outlined and presented in graphical form in Figure 13.6. It accounts for the nonlinear effects motivated in Figure 13.5. The four-step process is outlined as follows:

1. Specify the desired discrete-time filter requirements and attributes. Prewarp the discrete-time critical frequencies into their corresponding analog frequencies. Compute the order of the resulting analog prototype filter that meets or exceeds the analog filter's specifications.

2. Design an analog prototype filter $H_p(s)$ from the given prewarped continuous-time parameters.

3. Convert the analog prototype into an analog filter $H(s)$ using an appropriate frequency–frequency transform.

4. Design a digital filter $H(z)$ using a bilinear z-transform of $H(s)$, which automatically warps the frequency axis back to its original values.

While this method may initially seem to be complicated, it is actually simple to implement using a digital computer, as demonstrated by the following set of examples.

Example: Bilinear z-Transform Design

The synthesis of a discrete-time Butterworth filter that meets or exceeds the following specifications, using the bilinear z-transform, begins with specifying the discrete-time filter's design objectives. They include

- maximum passband attenuation = −3 dB,
- passband $f \in [0, 1]$ kHz,
- minimum stopband attenuation = −10 dB,
- stopband $f \in [2, 5]$ kHz, and
- sample frequency f_s = 10 kHz.

Step 1 requires that the digital filter frequencies (f_p = 1 kHz, f_a = 2 kHz, and f_s = 10 kHz) be prewarped to their analog counterparts. The prewarped continuous-time critical frequencies are given by passband and stopband (attenuation) pairs:

$$\varpi_p = 2\pi f_p \, / \, f_s = 2\pi(0.1) = 0.2\pi$$

$$\Omega_p = \frac{2}{T_s}\tan(\varpi_p / 2) = 20 \times 10^3 \tan(0.1\pi) = 6498 \text{ rad/s} \rightarrow 1.0345 \text{ kHz}$$

$$\varpi_a = 2\pi f_a \, / \, f_s = 2\pi(0.2) = 0.4\pi$$

$$\Omega_a = \frac{2}{T_s}\tan(\varpi_a / 2) = 20 \times 10^3 \tan(0.2\pi) = 14,531 \text{ rad/s} \rightarrow 2.312 \text{ kHz}$$

Step 2 designs an analog prototype filter $H_p(s)$. From these data, a second-order Butterworth prototype filter model is required, which has a −3-dB gain at $\Omega = 1$, a normalized transition bandwidth ratio of $k_d = \Omega_p/\Omega_a = 6498/14,531 = 0.447$, and a stopband gain bounded by −10 dB. The second-order Butterworth prototype filter has a transfer function given by

$$H_p(s) = \frac{1}{s^2 + 1.414s + 1}.$$

Step 3 defines an analog filter model. The prototype is redefined using a frequency–frequency transform. In this case, $H_p(s)$ is mapped into a final analog filter using a low-pass-to-low-pass frequency–frequency transform $s = s/\Omega_p$. The resulting analog filter becomes

$$H_a(s) = \frac{4.3 \times 10^7}{s^2 + 9.2 \times 10^3 s + 4.3 \times 10^7}.$$

Step 4 implements the bilinear z-transform. The final step is to apply the bilinear z-transform to $H(s)$ to obtain $H(z)$, which in the process warps the critical frequencies of $H_a(j\Omega)$, namely, Ω_p and Ω_a, into their original critical frequency locations ω_p and ω_a. That is,

$$H(z) = \frac{0.0676(z+1)^2}{z^2 - 1.142z + 0.412}.$$

The second-order digital filter meets the posted design specifications.

Example: Higher-Order Bilinear z-Transformed Filter

Design an elliptic low-pass IIR meeting or exceeding the following specifications, using the bilinear z-transform following the step-by-step procedure, beginning with listing the filter's specifications:

- sampling frequency f_s = 100 kHz;
- allowable passband deviation = 1 dB, passband range $f \in$ [0, 20] kHz;
- minimum stopband attenuation = 60 dB, stopband range $f \in$ [22.5, 50] kHz.

Step 1 requires that the discrete-time critical frequencies be prewarped into the corresponding analog frequencies that are

$$\Omega_p = \frac{2}{T_s} \tan(\varpi_p / 2) = 2 \times 10^5 \tan(0.2\pi) = 1.45308 \times 10^5 \text{ rad/s} \rightarrow 23.1265 \text{ kHz},$$

$$\Omega_a = \frac{2}{T_s} \tan(\varpi_a / 2) = 2 \times 10^5 \tan(0.225\pi) = 1.70816 \times 10^5 \text{ rad/s} \rightarrow 27.1862 \text{ kHz}.$$

Step 2 initiates a design of the analog prototype filter $H_p(s)$. These prewarped frequencies along with the passband and stopband gains result in an analog elliptic (Cauer) prototype filter of order $N = 8$ that is specified below:

$$H_p(s) = 0.000317 \frac{s^8 + 22.7s^6 + 100.3s^4 + 155.0s^2 + 78.7}{s^8 + 0.843s^7 + 2.324s^6 + 1.518s^5 +}$$
$$1.792s^4 + 0.838s^3 + 0.493s^2 + 0.133s + 0.028$$

Step 3 converts the analog prototype filter $H_p(s)$ model into the final analog filter $H(s)$ using low-pass-to-low-pass frequency–frequency transform, resulting in the model shown below:

$$H(s) = 0.000317 \frac{s^8 + 5.65 \times 10^{11} s^6 + 6.18 \times 10^{22} s^4 + 2.37 \times 10^{33} s^2 + 2.98 \times 10^{43}}{s^8 + 1.32 \times 10^5 s^7 + 5.77 \times 10^{10} s^6 + 5.935 \times 10^{15} s^5 + 1.10 \times 10^{21} s^4 +}$$
$$8.13 \times 10^{25} s^3 + 7.55 \times 10^{30} s^2 + 3.22 \times 10^{35} s + 1.06 \times 10^{40}$$

Step 4 implements the bilinear z-transform of $H(s)$ producing the $H(z)$ shown below. The bilinear z-transform warps the frequency axis that was previously prewarped,

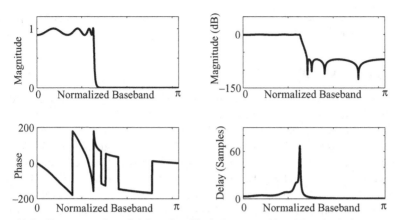

Figure 13.7   Response of an eighth-order elliptic low-pass filter. Magnitude frequency response (top left), magnitude frequency response in decibels (top right), phase response in degrees (bottom left), and group delay in samples (bottom right).

restoring the original critical frequencies. The result is the IIR shown below and meets the posted specifications:

$$H(z) = 0.00658 \frac{\begin{array}{c} z^8 + 1.726z^7 + 3.949z^6 + 4.936z^5 + \\ 5.923z^4 + 4.936z^3 + 3.949z^2 + 1.726z + 1 \end{array}}{\begin{array}{c} z^8 - 3.658z^7 + 7.495z^6 - 11.432z^5 + \\ 11.906z^4 - 8.996z^3 + 4.845z^2 - 1.711z + 0.317 \end{array}}.$$

The frequency response of the resulting elliptic IIR is shown in Figure 13.7. The passband deviation for the digital filter is 1 dB, and the minimum stopband attenuation is about 70 dB, which exceeds the 60-dB specification. The frequency response of the eighth-order elliptic filter is seen to exhibit the classic ripple in the pass- and stopband. Observe that most of the phase variability is concentrated in the pass- and transition band and early stopband. This is verified by viewing the group delay that indicates that a delay of about 60+ samples occurs at a transition band frequency.

MATLAB IIR DESIGN

Analog Butterworth, Chebyshev I and II, and elliptic filters can be designed using MATLAB `butter`, `cheby1`, `cheby2`, or `ellip` commands. MATLAB automatically implements the four-step design process. The process of developing a digital IIR filter using the bilinear z-transforms starts with defining the digital filter's performance specifications. These specifications are interpreted as analog filter requirements. The analog model is then converted into a discrete-time IIR filter using a bilinear z-transform. This process is well established and has been reduced to a computer program. MATLAB has a suite of programs that implement this process, and they are showcased below.

Example: MATLAB

Design a band-pass digital Butterworth, Chebyshev I and II, and elliptic band-pass IIR filter having a 1-dB passband ranging over $f \in [200, 300]$ Hz. The leading −50-dB stopband ranges over $f \in [0, 100]$ Hz. The lagging −50-dB stopband is defined over $f \in [400, 500]$ Hz. The sampling rate is 1000 Sa/s or the Nyquist frequency is 500 Hz. The designs include low-pass analog prototype filters of order $N = 5$ for Butterworth, $N = 4$ for Chebyshev I and II, and $N = 3$ for elliptic IIRs. These filter orders can be verified using MATLAB's `buttord`, `cheb1ord`, `cheb2ord`, and `ellipord` commands. The Nth-order low-pass filter will become a $2N$-order band-pass IIR when the low pass to band-pass frequency–frequency transform is performed. The Butterworth outcome is displayed in Figure 13.8, the Chebyshev I outcome in Figure 13.9, the Chebyshev II outcome in Figure 13.10, and the elliptic outcome in Figure 13.11. If the analog filter model $H(s)$ is known, then the MATLAB command `bilinear` can be used once f_s is specified.

The standard MATLAB-facilitated IIR design process first translates the IIR filter specifications into a filter order (`buttord`, `cheb1ord`, `cheb2ord`, and `ellipord`) and critical frequencies, and then uses this information to synthesize an IIR using the bilinear z-transform. The database for the first step is summarized in Table 13.1. The output of the order calculation (i.e., $[n, W_n]$) is then presented to the

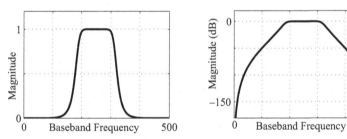

Figure 13.8 Tenth-order Butterworth IIR magnitude frequency response (left) and in decibels (right).

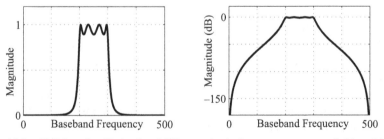

Figure 13.9 Eighth-order Chebyshev I IIR magnitude frequency response (left) and in decibels (right).

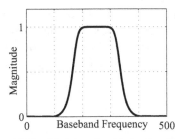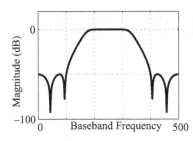

Figure 13.10   Eighth-order Chebyshev I IIR magnitude frequency response (left) and in decibels (right).

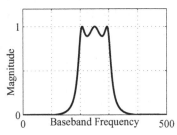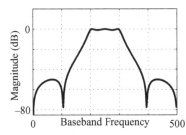

Figure 13.11   Sixth-order elliptic IIR magnitude frequency response (left) and in decibels (right).

TABLE 13.1.   Filter Order and Critical Frequencies (Frequencies Normalized by $f_s/2$) for Use with Butterord(W_p, W_a, R_p, R_s), Cheb1ord(W_p, W_a, R_p, R_s), Cheb2ord(W_p, W_a, R_p, R_s), **and** Ellipord(W_p, W_a, R_p, R_s) **Commands**

Class	Condition	Stopband	Passband
Low-pass	$W_p < W_s$	$[W_s, 1]$	$[0, W_p]$
High-pass	$W_p > W_s$	$[0, W_s]$	$[W_p, 1]$
Band-pass	$W_s(1) < W_p(1) < W_p(2) < W_s(2)$	$[0, W_s(1)]; [W_s(2), 1]$	$[W_p(1), W_p(2)]$
Band-stop	$W_p(1) < W_s(1) < W_s(2) < W_p(2)$	$[W_s(1), W_s(2)]$	$[0, W_p(1)]; [W_p(2), 1]$

R_p, passband deviation in decibels; R_s, minimum stopband attenuation in decibels; W_p, passband frequency; W_s, stopband frequency.

desired IIR module completing the command butter(n,Wn), cheby1(n,Wn), cheby2(n,Wn), or ellip(N,Wn), where N is the order of the analog prototype filter and W_n is generated by the order program. It should be noted that for low-pass and high-pass, the resulting IIR filter is of order N, but for a band-pass or band-stop designs, the outcome is a filter of order $2N$. This two-step method was used to produce the filters presented in Figures 13.8–13.11.

IMPULSE INVARIANCE VERSUS BILINEAR IIRS

An analog filter having a ramp impulse response $h(t) = tu(t)$ can be uniformly sampled and transformed using the standard and/or bilinear z-transform. While the system is unstable, it can nevertheless be used to explore the difference between the two design paradigms. For analysis purposes, assume that the sample rate is $f_s = 1$ Hz. The impulse invariant model $H(z)$ is defined in terms of the z-transform of $H(s) = 1/s^2$. Applying the standard and bilinear z-transform to $H(s)$ is, in either case, a straightforward mapping process:

$$H_1(z)|_{\text{standard}-z} = \frac{z}{z^2 - 2z + 1} = \frac{z^{-1}}{1 - 2z^{-1} + z^{-2}},$$

$$H_2(z)|_{\text{bilinear}-z} = 0.25\frac{z^2 + 2z + 1}{z^2 - 2z + 1} = 0.25\frac{1 + 2z^{-1} + z^{-2}}{1 - 2z^{-1} + z^{-2}}.$$

The short-term impulse responses can be computed using long division. In particular,

standard z-transform:

$$1 - 2z^{-1} + z^{-2} \overline{\big)}\ z^{-1} \quad \frac{1z^{-1} + 2z^{-2} + 3z^{-3} + \text{ramp time} - \text{series}}{}$$

$$\frac{z^{-1} - 2z^{-2} + z^{-3}}{\text{continue}}$$

or $h[k] = \{0, 1, 2, 3, \ldots\}$, which is immediately recognized to a ramp impulse response thereby reinforcing the impulse invariant claim of the standard z-transform. The bilinear z-transform results in

bilinear z-transform:

$$1 - 2z^{-1} + z^{-2} \overline{\big)}\ \frac{0.25 + 1z^{-1} + 2z^{-2} + \text{others}}{0.25 + 0.5z^{-1} + 0.25z^{-2}}$$

$$\frac{0.25 - 0.5z^{-1} + 0.25z^{-2}}{\text{continue}}$$

or $h[k] = \{0.25, 1, 2, \ldots\}$, which technically is not to be the impulse response of a ramp. That is, the bilinear z-transform produces an impulse response that does not agree with the value of $h(t)$ at the sample instances and, as a result, is not an impulse invariant transform.

Example: Impulse Invariant and Bilinear z-Transform

The second-order analog -1-dB Butterworth low-pass discrete-time filter having a 1-kHz passband previously studied is revisited. The analog filter has a transfer function $H(s)$ where

$$H(s) = \frac{7.757 \times 10^7}{s^2 + 1.2454 \times 10^4 s + 7.757 \times 10^7}.$$

The impulse invariant version of $H(s)$, sampled at 5000 Sa/s, is

$$H(z) = T_s \times \frac{3.3968 \times 10^3 z^1}{(z^2 - 0.1968z + 0.0828)} = \frac{0.67936z^1}{(z^2 - 0.1968z + 0.0828)}.$$

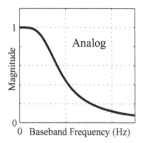 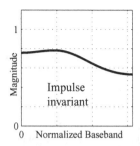 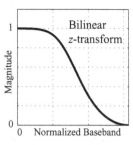

Figure 13.12 Butterworth analog filter response plotted out to 5 kHz (left). Impulse invariant filter response plotted out to the normalized Nyquist frequency of 2.5 kHz (middle); bilinear z-transform response plotted out to the normalized Nyquist frequency (right).

The bilinear z-transform version of $H(s)$, sampled at 5000 Sa/s, is

$$H(z) = \frac{0.2568 + 0.5135z^{-1}0.2568z^{-2}}{1.0 - 0.1485z^{-1} + 0.1755z^{-2}}.$$

The magnitude frequency response of the analog, impulse invariant, and bilinear z-transform filters are shown in Figure 13.12. It can be seen that the impulse invariant filter is a distorted version of the analog parent. The bilinear z-transform filter, however, meets the posted design requirements.

OPTIMIZATION

Classic filters approximate a piecewise constant frequency response. As a result, there are instances when a classic filter model does not exactly meet an applications need. In some cases, an approximate piecewise constant response is preferred. In such cases, it is desired to manipulate the response of a classic filter in an intelligent manner, distorting the piecewise constant frequency response profile into one that is similar but different. This can be accomplished by manipulating the pole and zero locations of the basic classic filter that "seeds" the final design. The movement of poles and zeros has a predictable effect on a filter's frequency response. A filter's frequency-dependent gain, for example, can be increased (peaked) by moving a pole or poles closer to the periphery of the unit circle. Frequency-dependent gains can be decreased (notched) by moving a zero or zeros closer to the periphery of the unit circle. These pole/zero movements can be iteratively preformed in software until a desired outcome is realized.

Example: Optimization

Suppose the passband of an *N*th-order IIR is to have a higher gain near the end of the passband compared with the DC gain value. This can be justified when the gain of the "upstream" filters and systems more aggressively attenuate the high-frequency

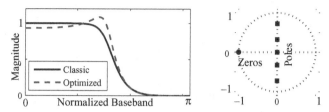

Figure 13.13 Magnitude frequency response of a classic Butterworth filter along with its optimized counterpart (left) and initial pole-zero distribution (right).

signal. By lifting the higher frequency gain with a modified classic filter, the end–end gain can be equalized. To illustrate, the magnitude frequency response of a fifth-order low pass Butterworth filter, shown in Figure 13.13, having a passband extending out to $f_p = 0.25 f_s$, is used to "seed" the design. The poles and zero of the classic IIR are shown in Figure 13.13 and have the following values:

$$\text{poles: } 0 \pm j0.7265; \ 0 \pm j0.3249; \ 0$$
$$\text{zeros: } -1, -1, -1, -1, -1.$$

Increasing the value of the complex-conjugate pole pairs closest to f_p, namely, $0 \pm j0.7265$, by 10% produces the bulge in the frequency response shown in Figure 13.13. By increasing or decreasing the pole scaling, slightly different effects can be achieved.

CHAPTER 14

STATE VARIABLE FILTER MODELS

STATE-DETERMINED SYSTEMS

The ability to form and manipulate transfer functions $H(z)$ is fundamentally important to the design and analysis of a linear time-invariant (LTI) system. A transfer function, however, only quantifies a filter's input–output behavior and does not describe the internal dynamics of a system as defined by the system's architecture. Architecture, in this instance, refers to how the basic system's building blocks (i.e., memory, adders, multipliers, and data paths) are assembled to implement a particular solution. What is desired is to develop an architecture-aware methodology that is capable of analyzing and auditing filter information, both internal and external. This is the role of state variables, a tool used in the study of filter architectures.

STATE VARIABLES

State variables represent the information residing in a filter's memory or registers. The future states of a state-determined system can be computed if

1. all past state variable values are known,
2. the mathematical relationship between the state variables are known, and
3. all future inputs are known.

In general, a multiple-input multiple-output (MIMO) discrete-time system consists of P-inputs, R-outputs, N-states, and a state variable representation model given by

$$\text{state equation: } \mathbf{x}[k+1] = \mathbf{A}[k]\mathbf{x}[k] + \mathbf{B}[k]\mathbf{u}[k], \qquad (14.1)$$

$$\text{initial conditions: } \mathbf{x}[0] = \mathbf{x}_0, \qquad (14.2)$$

$$\text{output equation: } \mathbf{y}[k] = \mathbf{C}^\mathrm{T}[k]\mathbf{x}[k] + \mathbf{D}[k]\mathbf{u}[k], \qquad (14.3)$$

where $\mathbf{A}[k]$ is an $N \times N$ matrix, $\mathbf{B}[k]$ is an $N \times P$ matrix, $\mathbf{C}[k]$ is an $N \times R$ matrix, $\mathbf{D}[k]$ is an $R \times P$ matrix, $\mathbf{u}[k]$ is an arbitrary $P \times 1$ input vector, $\mathbf{x}[k]$ is an $N \times 1$ state

Digital Filters: Principles and Applications with MATLAB, First Edition. Fred J. Taylor.
© 2012 by the Institute of Electrical and Electronics Engineers, Inc.
Published 2012 by John Wiley & Sons, Inc.

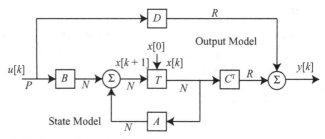

Figure 14.1 MIMO state-determined model.

vector, and $\mathbf{y}[k]$ is an $R \times 1$ output vector. Such a system can be represented by the state 4-tuple $[A[k], B[k], C[k], D[k]]$. If the discrete-time system is also an LTI system (i.e., constant coefficient), then the state 4-tuple consists of a collection of constant coefficient matrices $\{A, B, C, D\}$. A discrete-time state-determined system, based on Equations 14.1 through 14.3, is graphically interpreted in Figure 14.1. Note that the system's N state variables are stored in N shift registers. If an Nth-order system can be implemented using only N shift registers, the architecture is said to be canonic.

Many important discrete-time LTI filters are single-input single-output (SISO) systems, which can be modeled by the Nth-order difference equation of the form

$$a_0 y[k] + a_1 y[k-1] + \cdots + a_N y[k-N] = b_0 u[k] + b_1 u[k-1] + \cdots + b_N u[k-N]. \quad (14.4)$$

The discrete-time at-rest LTI system, defined by Equation 14.4, can also be expressed as a transfer function $H(z)$, where

$$
\begin{aligned}
H(z) &= \frac{b_0 + b_1 z^{-1} + \cdots + b_N z^{-N}}{a_0 + a_1 z^{-1} + \cdots + a_N z^{-N}} = \frac{b_0}{a_0} + \frac{(b_1 - b_0 a_1 / a_0)z^{-1} + \cdots + (b_N - b_0 a_N / a_0)z^{-N}}{a_0 + a_1 z^{-1} + \cdots + a_N z^{-N}} \\
&= \frac{b_0}{a_0} + \frac{c_N z^{-1} + \cdots + c_1 z^{-N}}{a_0 + a_1 z^{-1} + \cdots + a_N z^{-N}} = d_0 + C(z)\left(\frac{1}{D(z)}\right).
\end{aligned} \quad (14.5)
$$

If $a_0 = 1$, the system is said to be monic. The transfer function is seen to consist of three distinct subsystems. They are (1) a constant input–output path (d_0), (2) feed-forward paths ($C(z)$), and (3) feedback paths ($D(z)$).

SIMULATION

One of the attributes of a state variable system representation is its ability to support discrete-time simulation for a selected architecture. The system shown in Figure 14.1, consisting of coefficients, delays, and data paths, can be mapped to a state-space representation $\{A, b, c, d\}$. The state-space matrices and vectors can directly support a discrete-time simulation of the system under study. Specifically, given a state 4-tuple $\{A, b, c, d\}$, the total system response can be iteratively computed using the following:

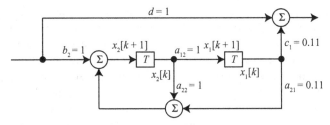

Figure 14.2   State-induced architecture.

Initialize: $[\mathbf{A}, \mathbf{b}, \mathbf{c}, d]$; x0, u

for $(i = 0, N)$

$$\mathbf{y}[k] = \mathbf{c}^{\mathrm{T}}[k]\mathbf{x}[k] + d[k]\mathbf{u}[k]$$

$$\mathbf{x}[k+1] = \mathbf{A}[k]\mathbf{x}[k] + \mathbf{b}[k]\mathbf{u}[k]$$

(14.6)

end

Example: Simulator

Suppose a specific second-order system is given by

$$A = \begin{bmatrix} 0 & 1 \\ 0.11 & 1 \end{bmatrix}; b = \begin{bmatrix} 0 \\ 1 \end{bmatrix}; c = \begin{bmatrix} 0.11 \\ 0 \end{bmatrix}; d = [1].$$

The architecture associated with this state description, namely, $\{A, b, c, d\}$, is shown in Figure 14.2.

Suppose the system is described by $\{A, b, c, d\}$, is initially at rest, and is presented with an input given as $u[k] = \delta[k] - \delta[k-1] - 0.11\delta[k-2]$. The state and output response, for $k = 0, 1, 2, 3, 4$, can be computed using MATLAB. The first five computed responses are $y[k] = \{1, 0, 0, 0, 0\}$, $x_1[k] = \{0, 0, 1, 0, 0\}$, and $x_2[k] = \{0, 1, 0, 0, 0\}$.

MATLAB SIMULATIONS

The act of simulating the response of an at-rest linear state-determined system to an arbitrary input can be reduced to an iterative computer program. In MATLAB, the command dlsim belongs to the MATLAB's Control Toolbox and can be used to support simulation studies. To illustrate, consider the design of a second-order Butterworth low-pass infinite impulse response (IIR), sampled at a 100 Sa/s rate, having a passband out to $f_p = (0.1)f_s = 10$ Hz. The filter's input is the sum of two sinusoids at frequencies $f = 0.2$ and 5 Hz, both being passband frequencies. The filter's state variable representation is given by

$$A = \begin{bmatrix} 0.277 & 0.415 \\ 0.415 & 0.865 \end{bmatrix}; b = \begin{bmatrix} 0.587 \\ 0.190 \end{bmatrix}; c^{\mathrm{T}} = \begin{bmatrix} 0.146 \\ 0.659 \end{bmatrix}; d = 067.$$

The resulting state-determined architecture is shown in Figure 14.3. Figure 14.4 displays the input signal and spectrum over $f \in [0, f_s]$ for an input containing two distinct passband frequencies. Figure 14.5 compares the input, output, and state trajectories. It would appear that even though the input and output trajectories are highly synergistic, the state trajectories seem to behave somewhat independently. This means the internal system behavior, in practice, must also be analyzed and quantified in detail and cannot be assumed. Figure 14.6 repeats the study reported in Figure 14.4 but does so using an input that has only one of its two frequencies in the filter's passband. Figure 14.7 addresses the study presented in Figure 14.5 based on the new signal.

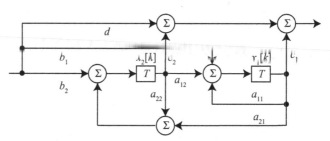

Figure 14.3   Derived architecture of a second-order Butterworth filter.

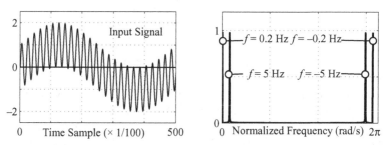

Figure 14.4   First 5 seconds of the multitone input signal having both signal frequencies in the filters passband (left). Normalized input spectrum ($f \in [0, 100]$ Hz) (right).

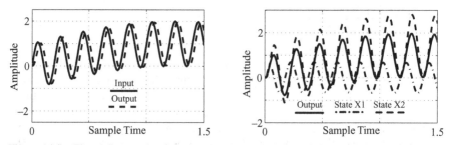

Figure 14.5   First 1.5 seconds of the input and output signals (left) and output, with states $x_1[k]$ and $x_2[k]$ outputs (right). Since the multitone input is entirely in the filter's passband, the output amplitude is similar to its input value. The output is, however, delayed relative to the input due to the filter's finite group delay. The state $x_1[k]$ appears to pass the higher input tone and attenuate the lower frequency component. The state $x_2[k]$, however, appears to have inherited multitone attributes.

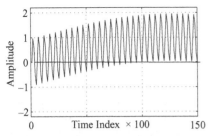
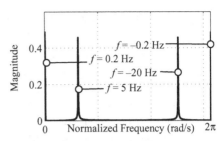

Figure 14.6   First 1.5 seconds of the multitone input signal having one signal component in the filter's passband, the other in the stopband (left) and input spectrum (right).

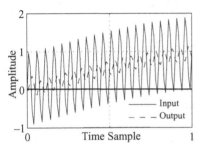
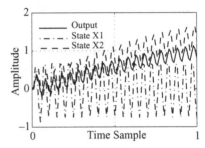

Figure 14.7   First 1.0 seconds of the multitone input and output signal (left). Output and states $x_1[k]$ and $x_2[k]$ are displayed (right). Since the multitone input is partially inside and outside the filter's passband, the output amplitude is different than the input amplitude. The output is again delayed relative to the input due to the filter's finite group delay. The state $x_1[k]$ again appears to contain the higher input tone and to attenuate the lower frequency component. The state $x_2[k]$, however, again appears to have inherited multitone attributes.

STATE VARIABLE MODEL

The SISO Nth-order LTI system, shown in Figure 14.1, can be modeled in terms of a state 4-tuple $\{A, b, c, d\}$ where $x[k]$ is the state N vector, $u[k]$ is the input, $y[k]$ is the output, A is an $N \times N$ state matrix, b is a $1 \times N$ input vector, c is a $1 \times N$ output vector, and d is a scalar. At some sample index $k = k_0$, the next state can be computed to be

$$x[k_0 + 1] = Ax[k_0] + bu[k_0]. \tag{14.7}$$

At sample instant $k = k_0 + 1$, the next state is given by

$$x[k_0 + 2] = Ax[k_0 + 1] + bu[k_0 + 1] = A^2 x[k_0] + Abu[k_0] + bu[k_0 + 1]. \tag{14.8}$$

Continuing this pattern, at sample $k = k_0 + n - 1$, the next state vector is given by

$$x[k_0 + n] = A^n x[k_0] + A^{n-1}bu[k_0] + A^{n-2}bu[k_0 + 1] + \dots$$
$$\dots + Abu[k_0 + n - 2] + bu[k_0 + n - 1]. \tag{14.9}$$

Equation 14.9 can be rearranged and expressed as

$$x[k] = A^{k-k_0}x[k_0] + \sum_{i=k_0}^{k-1} A^{k-i-1}bu[i].$$ (14.10)

Equation 14.10 is seen to consist of two parts, one defining the homogeneous solution, the other the inhomogeneous solution. The homogeneous solution is driven only by the initial conditions ($x[k_0]$), whereas the inhomogeneous solution is forced only by the external input $u[k]$. Upon substituting Equation 14.10 into the output equation (Eq. 14.3), the generalized output of the state-determined system is

$$y[k] = c^T x[k] + du[k] = \left(c^T A^{k-k_0} x[k_0] + c^T \sum_{i=k_0}^{k-1} A^{k-i-1}bu[k] + du[k] \right),$$ (14.11)

which represents a convolution process. The same result can be obtained using transform-domain techniques. Observe that the z-transform of Equations 14.1–14.3 are

$$zX(z) - zx_0 = AX(z) + bU(z),$$
$$Y(z) = c^T X(z) + dU(z).$$ (14.12)

If the system is at rest (i.e., $x_0 = 0$), then solving for $Y(z)$ one obtains

$$X(z) = (zI - A)^{-1}bU(z),$$
$$Y(z) = c^T(zI - A)^{-1}bU(z) + dU(z).$$ (14.13)

These results lead to the system's transfer function defined by $H(z) = Y(z)/U(z)$, or

$$H(z) = c^T(zI - A)^{-1}b + d.$$ (14.14)

Example: Transfer Function

An at-rest second-order LTI system is given by the state 4-tuple $\{A, b, c, d\}$:

$$A = \begin{bmatrix} 0 & 1 \\ 1 & 1 \end{bmatrix}; b = \begin{bmatrix} 0 \\ 1 \end{bmatrix}; c = \begin{bmatrix} 1 \\ 1 \end{bmatrix}; d = 1.$$

Using Equation 14.14, the transfer function can then be determined, and it is

$$H(z) = c^T(zI - A)^{-1}b + d = \frac{1}{z^2 - z - 1}[1 \ \ 1]\begin{bmatrix} z-1 & 1 \\ 1 & z \end{bmatrix}\begin{bmatrix} 0 \\ 1 \end{bmatrix} + 1$$

$$= \frac{z+1}{z^2 - z - 1} + 1 = \frac{z^2}{z^2 - z - 1}.$$

CHANGE OF BASIS

The architecture of an at-rest digital filter is defined by the state-determined 4-tuple $\{A, b, c, \mathrm{d}\}$, and the filter's transfer function is given by $H(z) = \mathbf{c}^T(z\mathbf{I} - \mathbf{A})^{-1}\mathbf{b} + \mathbf{d}$. Suppose that the state vector $x[k]$ is redefined to be a new state vector using the linear transform

$$\hat{\mathbf{x}}[k] = T\bar{\mathbf{x}}[k], \tag{14.15}$$

where T is a nonsingular $N \times N$ matrix. This corresponds to a change of basis. In the context of a digital filter, however, the linear transformation T corresponds to a redefining of the filter's architecture or wiring diagram. That is, the internal wiring diagram has been changed. Applying this transform to a state-determined system defined by $\{A, b, c, d\}$ will result in

$$
\begin{aligned}
\text{(state equation)} \quad \hat{\mathbf{x}}[k+1] &= T\bar{\mathbf{x}}[k+1] \\
&= TA\bar{\mathbf{x}}[k] + Tb\bar{u}[k] \\
&= TAT^{-1}\hat{\mathbf{x}}[k] + Tb\bar{u}[k] \\
&= \hat{A}\hat{x}[k] + \hat{b}\bar{u}[k]
\end{aligned} \tag{14.16}
$$

$$
\begin{aligned}
\text{(output equation)} \quad \bar{y}[k] &= c^T\bar{\mathbf{x}}[k] + d\bar{u}[k] \\
&= c^T T^{-1}\hat{\mathbf{x}}[k] + d\bar{u}[k] \\
&= \hat{c}^T\hat{\mathbf{x}}[k] + \hat{d}\bar{u}[k]
\end{aligned}
$$

where

$$
\begin{aligned}
\hat{A} &= TAT^{-1}; \quad A = T^{-1}\hat{A}T \\
\hat{b} &= Tb; \quad b = T^{-1}\hat{b} \\
\hat{c}^T &= c^T T^{-1}; \quad c^T = \hat{c}^T T \\
\hat{d} &= d.
\end{aligned} \tag{14.17}
$$

The nonsingular transform T creates a new architecture defined by the new state 4-tuple $\{\hat{A}, \hat{b}, \hat{c}, \hat{d}\}$. Using the matrix identity

$$(XYZ)^{-1} = Z^{-1}X^{-1}Y^{-1}, \tag{14.18}$$

it follows that the transfer function of the new system, under the mapping T, is given by

$$
\begin{aligned}
\hat{H}(z) &= \hat{c}^T\left(z\mathbf{I} - \hat{A}\right)^{-1}\hat{b} + \hat{d} = \mathbf{c}^T T^{-1}\left(z\mathbf{I} - TAT^{-1}\right)^{-1}T\mathbf{b} + \mathbf{d} \\
&= \hat{c}^T T^{-1}\left(T(z\mathbf{I} - A)T^{-1}\right)^{-1}T\mathbf{b} + \mathbf{d} = \mathbf{c}^T T^{-1}\left(T^{-1}\right)^{-1}(z\mathbf{I} - A)^{-1}(T)^{-1}T\mathbf{b} + \mathbf{d} \\
&= \mathbf{c}^T(z\mathbf{I} - A)^{-1}\mathbf{b} + \mathbf{d} = H(z).
\end{aligned} \tag{14.19}
$$

That is, the similarity transformation T leaves the input–output system behavior unaffected but does change the internal architecture specification of the system. The significance of this is huge in that it states that virtually any given transfer function $H(z)$ can be implemented with an infinite number of architectural choices.

MATLAB STATE SPACE

MATLAB contains a collection of state-determined commands that provide a means of connecting traditional transfer functions, pole-zero descriptions, state variable models, and second-order state variable models together. These mappings relate to the objects and elements shown below:

$$\text{Transfer function model: } H(z) = \frac{\sum_{i=0}^{N} b_i z^{-i}}{\sum_{i=0}^{N} a_i z^{-i}}$$

$$\text{Pole-zero model: } H(z) = k \frac{\prod_{i=1}^{N}\left(1-\beta_i z^{-1}\right)}{\prod_{i=1}^{N}\left(1-\alpha_i z^{-1}\right)} \tag{14.20}$$

$$\text{State variable model: } \{A, b, c, d\}\begin{cases} \mathbf{x}[k+1] = A\mathbf{x}[k] + b\nu[k] \\ y[k] = c^{\mathrm{T}}\mathbf{x}[k] + d\nu[k] \end{cases}$$

Many of the state-determined elements relate to a second-order system representation of systems having a transfer function:

$$H(z) = g \prod_{k=1}^{L} \frac{b_{0k} + b_{1k}z^{-1} + b_{2k}z^{-2}}{1 + a_{1k}z^{-1} + a_{2k}z^{-2}}. \tag{14.21}$$

These forms are interrelated using the following MATLAB commands:

SS2TF: The state-space to transfer function conversion command has a syntax [NUM,DEN] = SS2TF(A,B,C′,D,iu) and computes the input–output transfer function from the *iu*th input location (*iu* = 1 for SISO) where

$$H(z) = C^{\mathrm{T}}(zI - A)^{-1}B + D = \frac{Num(z)}{Den(z)}. \tag{14.22}$$

SS2ZP: The state-space to zero-pole conversion command has a syntax [Z,P,K] = SS2ZP(A,B,C,D,iu) and computes the input–output transfer function from the *iu*th input location (*iu* = 1 for SISO) where

$$H(z) = C^{\mathrm{T}}(zI - A)^{-1}B + D = k\frac{\prod(z - z_i)}{\prod(z - p_i)}. \tag{14.23}$$

Example: SS2TF/SSTPZ

Consider the second-order state-determined system:

$$A = \begin{bmatrix} 0 & 1 \\ 1 & 1 \end{bmatrix} ; b = \begin{bmatrix} 0 \\ 1 \end{bmatrix} ; c = \begin{bmatrix} 1 \\ 1 \end{bmatrix} ; d = 1.$$

The state 4-tuple $\{A, b, c, d\}$ defines the filter's transfer function to be given by

$$H(z) = \mathbf{c}^{\mathrm{T}}(z\mathbf{I} - \mathbf{A})^{-1}\mathbf{b} + d = \begin{bmatrix} 1 & 1 \end{bmatrix} \frac{\begin{bmatrix} z-1 & 1 \\ 1 & z \end{bmatrix}}{z^2 - z - 1} \begin{bmatrix} 0 \\ 1 \end{bmatrix} + 1 = \frac{z^2}{z^2 - z - 1}.$$

MATLAB can be used to map the state 4-tuple into a SISO transfer function using the following:

```
A=[0 1; 1 1]; b=[0 ; 1]; ct=[1 , 1]; d=1;

NUM, DEN]=SS2TF(A,b,ct,d,1);
```

which results in the transfer function $H(z) = z^2/(z^2 - z - 1)$. MATLAB can also map a state 4-tuple into a SISO transfer function in pole-zero form using $[Z,P,K] = $ SS2ZP(A,b,ct,d,1), which results in the transfer function

$$H(z) = \frac{(z-0)(z-0)}{(z+0.618)(z-0.618)}.$$

The reverse process can also be facilitated by MATLAB. Mapping from a state variable model to a transfer function is accomplished using the functions shown below:

TF2SS: The transfer function to state-space conversion command has a syntax [A,B,C,D] = TF2SS(NUM,DEN) and calculates the state-space representation given a transfer function using

$$H(z) = C^{\mathrm{T}}(zI - A)^{-1}B + D \Rightarrow \{A, b, c, d\}. \tag{14.24}$$

ZP2SS: The zero-pole to state-space conversion command has a syntax [A,B,C,D] = ZP2SS(Z,P,K) and calculates a state-space representation given a filter's pole-zero distribution:

$$H(z) = C^{T}(zI - A)^{-1}B + D = k\frac{\prod(z - z_i)}{\prod(z - p_i)}. \tag{14.25}$$

Example: TF2SS/PZ2SS

Given $H(z) = z^2/(z^2 - z - 1)$, the MATLAB script

```
NUM=[1 0 0]; DEN=[1 -1 -1];

[A,B,C,D]=TF2SS(NUM,DEN)
```

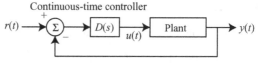

Figure 14.8 Digital controller.

produces a state 4-tuple $\left\{ A = \begin{bmatrix} 0 & 1 \\ 1 & 1 \end{bmatrix} : b = \begin{bmatrix} 0 \\ 1 \end{bmatrix} ; c = \begin{bmatrix} 1 \\ 1 \end{bmatrix} ; d = 1 \right\}$. Given the transfer function $H(z)$ in pole-zero form, $H(z) = (z - 0)(z - 0)/(z + 0.618)(z - 1.618)$, MATLAB can compute the state 4-tuple using

```
z=[0 0];  p=[-0.618 1.618];  k 1;

[A,B,C,D]=ZP2SS(Z,P,K):
```

producing $\left\{ A = \begin{bmatrix} 0 & 1 \\ 1 & 1 \end{bmatrix} : b = \begin{bmatrix} 0 \\ 1 \end{bmatrix} ; c = \begin{bmatrix} 1 \\ 1 \end{bmatrix} ; d = 1 \right\}.$

Example: Automatic Control

One of the principle signal processing domains is automatic controls. Controls have a natural affinity to state variable modeling and are often the principal analysis tool. The system shown in Figure 14.8 is that of a typical continuous-time feedback control system. The continuous-time controller can be replaced by an equivalent digital controller. The digital controller performs the same control task as the continuous-time controller with the basic difference being that it is a digital system. There is a MATLAB Control Toolkit function called c2dm that converts a given continuous system (either in transfer function or state-space form) into a discrete system model using a zero-order hold operation. The basic c2dm command has a format:

```
[numDz,denDz]=c2dm(num,den,Ts,'zoh')%transfer function

[F,G,H,J]=c2dm(A,B,C,D,Ts,'zoh')%state variables
```

As a guide, a control system's sampling time (T_s in sample per second) should be smaller than $1/(30*BW)$, where BW is the system's closed-loop bandwidth frequency (i.e., oversampling). Consider the continuous-time transfer function of a second-order mechanical system with feedback described by the transfer function $H(s) = 1/(Ms^2 + bs + k)$ where $M = 1$ kg, $b = 10$, $k = 20$ N/m, and $F(s) = 1$. Assuming the closed-loop bandwidth frequency is greater than 1 rad/s, choosing the sampling time (T_s) equal to $1/100$ second ($f_s = 100$ Sa/s) and parameters shown below produces the following outcome:

```
num=[1]; den=[M b k]; Ts=1/100;

[numDz,denDz]=c2dm(num,den,Ts,'zoh')

numDz = 1.0e-04 % k; 0, 0.4837, 0.4678 % N(z)
```

```
denDz = 1.0000,  -1.9029,  0.9048  %  D(z)

num=[0.4837,  0.4678];

den=[1 -1.9029 0.09048];

[A,b,c,d]=tf2ss(0.0001*num,  den);
```

MATLAB produces the discrete time controller replacement transfer function

$$H(z) = \frac{0.0001(0.4837z + 0.4678)}{(z^2 - 1.9029z + 0.0946)}$$

that has a state model illustrated in Figure 14.9 and given by

$$A = \begin{bmatrix} 1.902 & -0.090 \\ 1.000 & 0.000 \end{bmatrix}; b = \begin{bmatrix} 1 \\ 0 \end{bmatrix}; c^{\mathrm{T}} = \begin{bmatrix} 0.483 \\ 0.467 \end{bmatrix} \times 10^{-4}; d = 0.000.$$

TRANSPOSE SYSTEMS

The state variable 4-tuple $\{A, b, c, d\}$ defines a unique architecture. Another architecture, called a transpose architecture, can be derived from $\{A, b, c, d\}$. The system exemplified by the first-order system shown in Figure 14.10 can be reengineered as the transpose system also shown in Figure 14.10. The reengineering process is developed below:

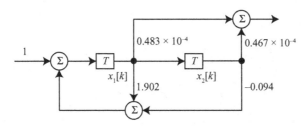

Figure 14.9   State model for the second-order system $H(z)$.

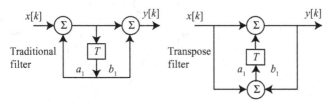

Figure 14.10   Traditional architecture (left) and transposed architecture (right) filters.

- Reverse the direction of flow in all branches.
- Reverse the input and output ports.
- All nodes become summers, and all summers become nodes.

The transpose of the state-determined system $\{A, b, c, d\}$ is characterized by the 4-tuple $\{A^T, c^T, b^T, d^T\}$. The resulting transfer function is

$$H(z) = d + c^T(z\mathbf{I} - A)^{-1}b = d^T + b^T(z\mathbf{I} - A^T)^{-1}(c^T)^T = d + b^T(z\mathbf{I} - A^T)^{-1}c. \quad (14.26)$$

The input–output transfer functions of the original and transpose systems are produced in Equation 14.26 and are equal. However, they have different internal wiring instructions.

Example: Transpose

Derive the state variable representation of a state-determined second-order Butterworth low-pass filter having a passband cutoff frequency of $f = 0.1f_s$. Convert the resulting filter's state 4-tuple $\{A, b, c, d\}$ into its transposed form $\{A^T, c^T, b^T, d^T\}$ using the MATLAB code shown below. The transfer function is $H(z) = (0.06746z^2 + 0.1349z + 0.06746)/(z^2 - 1.143z + 0.4128)$ and has a transposed architecture given by $\{A^T, c^T, b^T, d^T\}$. The MATLAB-produced transfer function is $H(z) = (0.06746z^2 + 0.1349z + 0.06746)/(z^2 - 1.143z + 0.4128)$, which is the same as the original, but synthesized from the transposed architecture given by $\{A^T, c^T, b^T, d^T\}$. When the two architectures are compared, however, they differ as shown in Figure 14.11. The outcomes can be explored using the following:

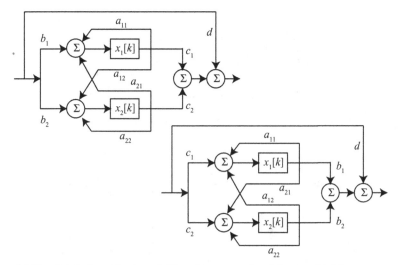

Figure 14.11 Original architecture (left) and transposed architecture (right).

```
[A,B,C,D]=butter(2,0.2);

[n,d]=ss2tf(A,B,C,D);  tf(n,d);  % original

[nt,dt]=ss2tf(A',C',B',D);tf(nt,dt);%transponse
```

MATLAB STATE-SPACE ARCHITECTURAL STRUCTURES

MATLAB provides a collection of objects that map a transfer function into a state-space architecture. The MATLAB Signal Processing Toolbox operator dfilt is used to describe a discrete-time filter and has the syntax H_d = dfilt.structure (input1,…). The function dfilt returns a discrete-time filter having a selected architecture or structure. The state-variable-specified filter structure is dfilt. statespace where H_d = dfilt.statespace(A,B,C,D) returns a discrete-time state-space filter defined in terms of the state 4-tuple $\{A, B, C, D\}$ that is graphically interpreted in Figure 14.12.

Example: MATLAB's State-Space Model

Design a sixth-order Chebyshev II low-pass filter having a −40-dB or better stop-band, and a stopband critical frequency of $f = 0.125f_s$. Implement the filter using a state variable model, test the filter's stability, and conduct filtering studies. The MATLAB-orchestrated design and analysis outcome is showcased in Figure 14.13.

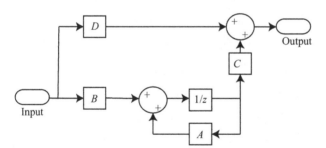

Figure 14.12   MATLAB's state variable system model displayed in MATLAB style (dfilt. statespace).

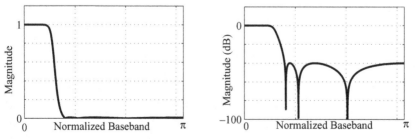

Figure 14.13   Low-pass Chebyshev II state variable filter's magnitude frequency response (left) and in decibels (right).

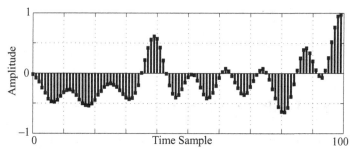

Figure 14.14   Output to the low-pass Chebyshev II state variable filter. The output time series is seen to be a smoothed version of a random input signal.

The MATLAB script shown below computes the filter's state 4-tuple for the initial states set to zero. The filter's bounded-input bounded-output (BIBO) stability is MATLAB-verified. The system is also tested using MATLAB and a zero-mean random input $n[k]$ with the response shown in Figure 14.14.

```
[n,d]=cheby2(6,40,0.25);  [A,b,c,d]=tf2ss(n,d);   % State
variable model

Hd=dfilt.statespace(A,b,c,d);   % Implement using MATLAB's
state space tool

get(Hd)  % report state variable model

get(Hd,'A')  % manually retrieve A

Hs=Hd.states   % Check that the initial states are zero

isstable(Hd)   % Is the filter BIBO stable?

x=randn(1000,1):   % input noise

y=filter(Hd,x);  % convolution
```

CHAPTER 15

DIGITAL FILTER ARCHITECTURE

FILTER ARCHITECTURE

A number of digital filter design procedures exist that translate a set of frequency-domain specifications into a transfer function $H(z)$. An infinite impulse response (IIR) filter, for example, can be based on a transfer function $H(z)$ defined in terms of polynomials or poles and zeros as suggested in Equation 15.1:

$$H(z) = \frac{N(z)}{D(z)} = K\frac{\displaystyle\sum_{m=0}^{M} b_m z^{-m}}{1+\displaystyle\sum_{m=1}^{N} a_m z^{-m}} = K\frac{\displaystyle\prod_{i=0}^{N-1}(z-z_i)}{\displaystyle\prod_{i=0}^{N-1}(z-\lambda_i)}. \tag{15.1}$$

Once the filter is designed, an implementing architecture needs to be selected. It should be appreciated that there are many architectures that are currently in common use. Examples of classic IIR architectures include direct I and II, cascade, parallel, ladder/lattice, and normal, to name a few. These architectures represent various trade-offs between maximum real-time speed, complexity, and precision. One of the most basic and popular architectures is called the direct form.

DIRECT I AND II ARCHITECTURES

Architectures can be viewed as a filter's wiring diagram. What differentiates one specific architecture from another are the details of this wiring diagram, sometimes called a netlist. The simplest architectural representation for a system having a transfer function $H(z)$ is the direct I form (architecture) shown in Figure 15.1. The direct I architecture, however, is noncanonic, requiring $2N$ shift registers to implement an Nth-order filter. The canonic direct II model is a refinement of the direct I architecture and is also shown in Figure 15.1. The canonic direct II structure needs only N shift registers to implement an Nth-order IIR. The advantage of the direct II over a direct I design is reduced complexity. The direct II architecture permits a

---

Digital Filters: Principles and Applications with MATLAB, First Edition. Fred J. Taylor.
© 2012 by the Institute of Electrical and Electronics Engineers, Inc.
Published 2012 by John Wiley & Sons, Inc.

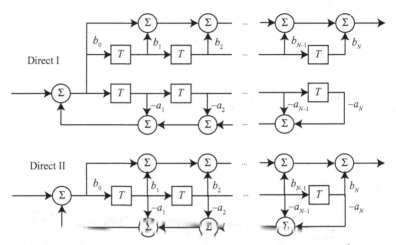

Figure 15.1   Direct I (top) and II (bottom) architectures. Coefficients a_i and b_i correspond to the multipliers of z^{-i} in $D(z)$ and $N(z)$. State $x_1[k]$ is closest to the output; state $x_N[k]$ is closest to the input.

tentative state assignment that locates the ith state variable $x_i[k]$, $i \in [1, N]$, at the ith shift register. This action is consistent with the knowledge that the states represent information, and information is stored in the filter's registers. The displayed direct II state assignment $x_N[k]$ to $x_1[k]$, shown in Figure 15.1, is stored in a serial chain of registers. Unfortunately, this model provides no convenient way to express the output $y[k]$ as a linear combination of the current states $x_i[k]$, $i \in [1, N]$, and the input $u[k]$. Refer to Figure 15.1 and note that $x_N[k + 1]$ (b_0 path) is not an element of the state N-vector $\mathbf{x}[k] = [x_1[k], \ldots, x_N[k]]^T$. This problem can be mitigated by slightly modifying the direct II form.

Consider representing $H(z)$ as

$$H(z) = \frac{N(z)}{D(z)} = \frac{b_0 + b_1 z^{-1} + \cdots + b_N z^{-N}}{a_0 + a_1 z^{-1} + \cdots + a_N z^{-N}} =$$

$$= \frac{b_0}{a_0} + \frac{(b_1 - b_0 a_1 / a_0)z^{-1} + \cdots + (b_N - b_0 a_N / a_0)z^{-N}}{a_0 + a_1 z^{-1} + \cdots + a_N z^{-N}} \qquad (15.2)$$

$$= \frac{b_0}{a_0} + \frac{c_N z^{-1} + \cdots + c_1 z^{-N}}{a_0 + a_1 z^{-1} + \cdots + a_N z^{-N}} = d_0 + C(z)\left(\frac{1}{D(z)}\right),$$

along with the following state assignment:

$$\mathbf{x}[k] = \begin{bmatrix} x_1[k] \\ x_2[k] \\ \vdots \\ x_N[k] \end{bmatrix} = \begin{bmatrix} x[k - N] \\ x[k - N + 1] \\ \vdots \\ x[k] \end{bmatrix}, \qquad (15.3)$$

where $x_i[k]$ is called the ith state variable and $\mathbf{x}[k]$ is the N-state vector of states $x_i[k]$ at clock instance k. This assignment allows the next state, $\mathbf{x}[k + 1]$, to be defined as

$$\mathbf{x}[k+1] = \begin{bmatrix} x_1[k+1] \\ x_2[k+1] \\ \vdots \\ x_N[k+1] \end{bmatrix} = \begin{bmatrix} x_2[k] \\ x_3[k] \\ \vdots \\ -a_N x_1[k] - a_{N-1} x_2[k] - \cdots - a_2 x_{N-1}[k] - a_1 x_N[k] + u[k] \end{bmatrix}.$$

(15.4)

This, in turn, enables a direct II architecture to be defined by $\mathbf{x}[k+1] = A\mathbf{x}[k] + \mathbf{b}u[k]$ where the $N \times N$ coefficient state matrix A and $N \times 1$ input vector $\mathbf{b}$ are defined to be

$$A = \begin{bmatrix} 0 & 1 & 0 & \cdots & 0 \\ 0 & 0 & 1 & \cdots & 0 \\ \vdots & \vdots & \vdots & \ddots & \vdots \\ 0 & 0 & 0 & \cdots & 1 \\ -a_N & -a_{N-1} & -a_{N-2} & \cdots & -a_1 \end{bmatrix}; \mathbf{b} = \begin{bmatrix} 0 \\ \vdots \\ 0 \\ 1 \end{bmatrix}.$$

(15.5)

The $[i, j]$ element of A defines the path gain between state $x_j[k]$ and $x_i[k+1]$. The output, or output state equation, is given by $y[k] = \mathbf{c}^T \mathbf{x}[k] + d_0 u[k]$. According to Equation 15.2, for a monic system (i.e., $a_0 = 1$),

$$\begin{aligned} \mathbf{c}^T &= \begin{pmatrix} b_N - b_0 a_N & b_{N-1} - b_0 a_{N-1} & \cdots & b_1 - b_0 a_1 \end{pmatrix} = \begin{pmatrix} c_1 & c_2 & \cdots & c_N \end{pmatrix} \\ d_0 &= b_0 \end{aligned}$$

(15.6)

where $\mathbf{c}$ is an N vector and d_0 is a scalar. It can be seen that the modified direct II architecture allows the output $y[k]$ to be constructed as a linear combination of the current states $x_i[k]$ and input $u[k]$. The result is a system having a state representation $\{A, \mathbf{b}, \mathbf{c}, d\}$. This modified direct II architecture is shown in Figure 15.2. However, due to its popularity, it is often referred to as simply the direct II architecture.

Example: Direct II Architecture

Consider, for example, a third-order monic digital filter having a transfer function given by

$$H(z) = \frac{1 - 0.5z^{-1} - 0.315z^{-2} - 0.185z^{-3}}{1 - 0.5z^{-1} + 0.5z^{-2} - 0.25z^{-3}} = 1 + \frac{-0.815z^{-2} + 0.065z^{-3}}{1 - 0.5z^{-1} + 0.5z^{-2} - 0.25z^{-3}}.$$

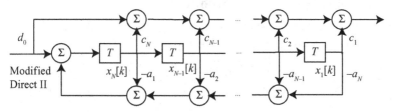

Figure 15.2   Modified direct II architecture (often called direct II).

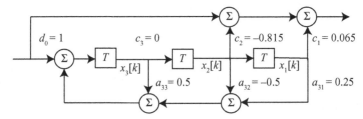

Figure 15.3 Third-order state-determined digital filter reduced to a direct II architecture showing traditional state assignments.

Using Equations 15.5 and 15.6, the modified direct II system can be expressed as the state 4-tuple:

$$
\mathbf{A} = \begin{pmatrix} 0 & 1 & 0 \\ 0 & 0 & 1 \\ 0.25 & -0.5 & 0.5 \end{pmatrix}; \mathbf{b} = \begin{pmatrix} 0 \\ 0 \\ 1 \end{pmatrix}; \mathbf{c} = \begin{pmatrix} 0.065 \\ -0.815 \\ 0.0 \end{pmatrix}; d = 1
$$

as shown in Figure 15.3.

DIRECT I AND II MATLAB IIR SUPPORT

MATLAB has extensive state variable direct I and II support. MATLAB's state variable programs, however, makes their state assignments in a reverse order from those normally found in signal processing literature and textbooks. Specifically, the traditional state located at $x_i[k]$ appears at state locations $x_{N-i}[k]$ in MATLAB. The conversion of a transfer function $H(z)$ into a direct II architecture can be performed using the MATLAB programs TF2SS or ZP2SS. For single-input single-output (SISO) IIR filter applications, they are briefly summarized below:

> TF2SS: Transfer function to state-space conversion given by [A,B,C,D] = TF2SS(NUM,DEN).

> ZP2SS: Zero-pole to state-space conversion given by [A,B,C,D] = ZP2SS (Z,P,K).

The resulting state 4-tuples {$\mathbf{A}, \mathbf{b}, \mathbf{c}, d$} is that of a direct II architecture in MATLAB style (i.e., reversed state order).

Example: MATLAB Direct II Architecture

To illustrate MATLAB's reordering of states, consider the traditional direct II architecture presented in Figure 15.3 and given by

$$
H(z) = \frac{1 - 0.5z^{-1} - 0.315z^{-2} - 0.185z^{-3}}{1 - 0.5z^{-1} + 0.5z^{-2} - 0.25z^{-3}} = 1 + \frac{-0.815z^{-2} + 0.065z^{-3}}{1 - 0.5z^{-1} + 0.5z^{-2} - 0.25z^{-3}}.
$$

MATLAB can map $H(z)$ into a direct II filter in MATLAB style. The MATLAB direct II filter can be obtained using the TF2SS command, producing

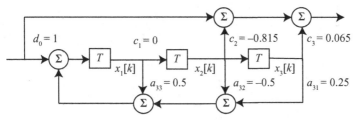

Figure 15.4   Third-order state-determined digital filter reduced to a direct II architecture showing MATLAB style state assignments.

```
n=[1 -.5 -.315 -.0185];  d=[1 -.5 .5 -.25];

[A,B,C,D]=TF2SS(n,d)
```

The outcome is

$$A = \begin{bmatrix} 0.5 & -0.5 & 0.25 \\ 1 & 0 & 0 \\ 0 & 1 & 0 \end{bmatrix}; \quad b = \begin{bmatrix} 1 \\ 0 \\ 0 \end{bmatrix}; \quad c = \begin{bmatrix} 0 \\ -0.8150 \\ 0.0650 \end{bmatrix}; \quad d = 1$$

and illustrated in Figure 15.4. Comparing the architectures shown in Figures 15.3 and 15.4, it can be seen that the architectures are identical except for the state ordering.

Example: High-Order Direct II Architecture

Consider the fifth-order digital filter, having a transfer function

$$
\begin{aligned}
H(z) &= 0.108 \frac{1 + 5z^{-1} + 10z^{-2} + 10z^{-3} + 5z^{-4} + z^{-5}}{1 + 0.98z^{-1} + 0.97z^{-2} + 0.38z^{-3} + 0.11z^{-4} + 0.01z^{-5}} \\
&= 0.108 \left(1 + \frac{(4.01)z^{-1} + (9.025)z^{-2} + (9.61)z^{-3} + (4.88)z^{-4} + (0.98)z^{-5}}{1 + 0.98z^{-1} + 0.97z^{-2} + 0.38z^{-3} + 0.11z^{-4} + 0.01z^{-5}} \right) \\
&= k \left(d + \frac{C(z)}{D(z)} \right).
\end{aligned}
$$

From Equation 15.2, $H(z) = k(d + C(z)/D(z))$ and consists of two parts. The first part is an unscaled scaled transfer function $H'(z) = (d + C(z)/D(z))$, which can be implemented as a direct II architecture. The state 4-tuple of the direct II filter, in MATLAB style, is

$$A = \begin{bmatrix} -0.986 & -0.974 & -0.387 & -0.111 & -9.011 \\ 1.0 & 0.0 & 0.0 & 0.0 & 0.0 \\ 0.0 & 1.0 & 0.0 & 0.0 & 0.0 \\ 0.0 & 0.0 & 1.0 & 0.0 & 0.0 \\ 0.0 & 0.0 & 0.0 & 1.0 & 0.0 \end{bmatrix}; \quad b = \begin{bmatrix} 1 \\ 0 \\ 0 \\ 0 \\ 0 \end{bmatrix}; \quad c = \begin{bmatrix} 0.433 \\ 0.974 \\ 1.038 \\ 0.528 \\ 0.106 \end{bmatrix}; \quad d = 0.108.$$

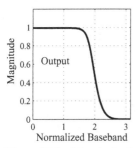 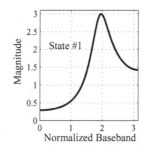 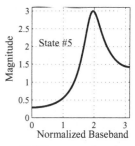

Figure 15.5 Magnitude frequency response of the fifth-order direct II IIR measured at the output $y[k]$ (left), state locations $x_1[k]$ (middle), and $x_5[k]$ (right).

The second part is the application of gain factor $k = 0.108$ to the first part transfer function $H'(z) = (d + C(z)/D(z))$ to form $H(z) = kH'(z)$. This can be achieved by scaling the input to form $u'(t) = ku(t)$. The analysis of the filter requires that the output magnitude frequency response, as well as those of the states be quantified. To compute the individual state responses, first realize that states 1 through 5 are connected by a first-in first-out (FIFO) shift-register chain. Therefore, what happens to MATLAB state $x_1[k]$ happens to state $x_5[k]$ five samples later. Next, recognize that state $x_i[k]$ can be isolated in the output by making the following modifications to c and d:

$$c^T = \begin{bmatrix} 1 & 0 & 0 & 0 & 0 \end{bmatrix}; d = 0; y[k] = x_1[k]$$

$$c^T = \begin{bmatrix} 0 & 1 & 0 & 0 & 0 \end{bmatrix}; d = 0; y[k] = x_2[k]$$

$$c^T = \begin{bmatrix} 0 & 0 & 1 & 0 & 0 \end{bmatrix}; d = 0; y[k] = x_3[k] .$$

$$c^T = \begin{bmatrix} 0 & 0 & 0 & 1 & 0 \end{bmatrix}; d = 0; y[k] = x_4[k]$$

$$c^T = \begin{bmatrix} 0 & 0 & 0 & 0 & 1 \end{bmatrix}; d = 0; y[k] = x_5[k]$$

Figure 15.5 displays the output frequency response along with those of the isolated states $x_1[k]$ and $x_5[k]$.

Since $x_1[k] = x_2[k-1] = x_3[k-2] = x_4[k-3] = x_5[k-4]$, the magnitude frequency responses of $x_1[k]$ and $x_5[k]$ are identical. It can be observed that the magnitude frequency responses of the state solutions have a high local gain, which is about three times larger than that found in the natural output $y[k]$.

MATLAB DIRECT I AND II STRUCTURES

MATLAB also provides a collection of objects that map a transfer function into a set of direct I and direct II architectures. The MATLAB Signal Processing Toolbox command dfilt is used to describe a discrete-time filter with a specific architecture or structure using the syntax Hd = dfilt.structure(input1, ...). Each structure takes one or more inputs where the direct filter structures are as follows:

dfilt.df1   direct-form I,

dfilt.df1t   direct-form I transposed,

dfilt.df2   direct-form II,

dfilt.df2t   direct-form II transposed.

The dfilt function forms a family of MATLAB direct filters that are summarized in Figure 15.1. Transposed versions of direct filter architectures are developed in Chapter 14. It should be noted that the df1 and df1t options are noncanonic, while the df2 and df2t are canonic.

Example: MATLAB Direct IIR Structure

Design a sixth-order Chebyshev II low-pass filter having a -40 dB or better stop-band, and a stopband critical frequency of $f = 0.125f_s$. The IIR can be implemented using a MATLAB direct I (df1), direct I transposed (df1t), direct II (df2), and direct II transposed (df2t) architecture. The evaluation procedure for a direct II is shown below:

```
[n,d]=cheby2(6,40,0.25);

Hd=dfilt.df2(n,d);  %   Direct II

get(Hd)

Numerator: [0.0164 -0.0243 0.0375 -0.0292 0.0375 -0.0243
0.0164]

Denominator: [1 -3.5779 5.6767 -4.9900 2.5467 -0.7101
0.0846]
```

It can be seen that the IIR analysis reports a transfer function numerator and denominator, which can be displayed as a state-determined direct II for using the TF2SS command.

CASCADE ARCHITECTURE

One of the most popular IIR structures found in common use today is called the cascade architecture. The basic Q-stage cascade architecture is shown in Figure 15.6 and can be represented with a transfer function having the form

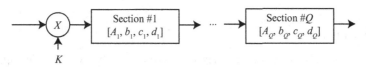

Figure 15.6   Cascaded architecture consisting of Q subfilters.

$$H(z) = \frac{N(z)}{D(z)} = G\prod_{i=1}^{Q} H_i(z); \quad \sum_{i=1}^{Q} \text{order}(H_i(z)) = N. \tag{15.7}$$

The ith subsystem, denoted $H_i(z)$, is either a first- or second-order filter section having only real coefficients. Each low-order subfilter can also be represented in terms of a state model $\{A_i, b_i, c_i, d_i\}$ of order $N_i \in [1, 2]$. A cascade architecture, as the name implies, presents the output of the ith subsystem to the input of its successor. The design rules for the creation of first- and second-order subfilters are well established and are presented in the next section.

FIRST- AND SECOND-ORDER SUBFILTERS

A transfer function $H(z) = N(z)/D(z)$ possesses many possible factorizations. One obvious form is based on the filter's pole-zero distribution. Specifically,

$$H_i(z) = \frac{N(z)}{D(z)} = \frac{b_0 + b_1 z^{-1} + \ldots + b_{N_i} z^{-N_i}}{a_0 + a_1 z^{-1} + \ldots + a_{N_i} z^{-N_i}} = \frac{\displaystyle\prod_{j=1}^{N_i}(z - z_j)}{\displaystyle\prod_{j=1}^{N_i}(z - p_j)}. \tag{15.8}$$

where $H_i(z)$ is an N_i-order IIR, $N_i < N$. Generally, for cascade designs, the real or complex conjugate poles and zeros are combined to create first- or second-order subsystems having only real coefficients. These first- and second-order systems can provide a basic design framework for cascade as well as other filter architectures.

First-Order IIR

First-order subsystems occur when one or more of the system pole or poles are real. Suppose p_i is real; then, a first-order IIR biquadratic or direct II or subfilter can be constructed, where

$$H_i(z) = \frac{q_{i0} + q_{i1}z^{-1}}{1 - p_i z^{-1}} \text{ biquadartic (biquad)},$$

$$H_i(z) = q_{i0} + \frac{r_i z^{-1}}{1 - p_i z^{-1}}; \; r_i = q_{i1} + p_i q_{i0} \text{ direct II.} \tag{15.9}$$

The details of the two first-order architectural choices are shown in Figure 15.7.

Second-Order IIR

Second-order subsystems occur when one or more of the system poles appear in complex-conjugate pairs. Suppose p_i and p_i^* are complex-conjugate poles; then, second-order IIR biquadratic or direct II or subfilter can be constructed, where

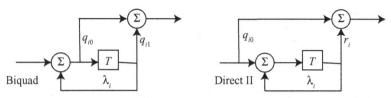

Figure 15.7 Implementation of a first-order section as a biquad $H_i(z) = (q_{i0} + q_{i1}z^{-1})/(1 - \lambda_i z^{-1})$ (left) or direct II section $H_i(z) = q_{i0} + (r_i z^{-1})/(1 - \lambda_i z^{-1})$ (right).

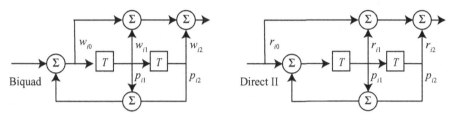

Figure 15.8 Implementation of a second-order section as a biquad $H_i(z) = (w_{i0} + w_{i1}z^{-1} + w_{i2}z^{-2})/(1 - p_{i1}z^{-1} - p_{i2}z^{-2})$ (left) or direct II section $H_i(z) = r_{i0} + (r_{i1}z^{-1} + r_{i2}z^{-2})/(1 - p_{i1}z^{-1} - p_{i2}z^{-2})$ (right).

$$H_i(z) = \frac{w_{i0} + w_{i1}z^{-1} + w_{i2}z^{-2}}{1 - v_{i1}z^{-1} - v_{i2}z^{-2}}; (z - p_i)(z - p_i^*) = z^2 - v_{i1}z - v_{i2}; \text{ biquadartic (biquad)}$$

$$H_i(z) = d_{i0} + \frac{r_{i1}z^{-1} + r_{i2}z^{-2}}{1 - v_{i1}z^{-1} - v_{i2}z^{-2}}; \begin{array}{c} (z - p_i)(z - p_i^*) = z^2 - v_{i1}z - v_{i2} \\ d_{i0} = w_{i0}; r_{i1} = w_{i1} + v_{i1}w_{i0}; r_{i2} = w_{i2} + v_{i2}w_{i0} \end{array} ; \text{ direct II.}$$

$$(15.10)$$

The details of the two second-order architectural choices are shown in Figure 15.8.

Both first- and second-order filters are defined in terms of real multiplications, which, compared with complex arithmetic, are very efficient. A logical question to raise is how to pair the first- and/or second-order zeros and poles from a field of M possible zeros and N poles. For each pairing, a different subfilter configuration is realized having a unique frequency response and maximum gain. As a general rule, zeros are paired with the closest poles. This proximity pairing strategy will generally result in a filter design having filter sections possessing the smallest maximal gain across the baseband, spreading the gain requirements more equitably across all subfilters. Other pairing strategies can result in a few subsystems having excessively large dynamic range, which can reduce overall system precision. To illustrate, consider a simple second-order IIR having poles at $z = 0.9$ and -0.9, with zeros located at $z = 0.5$ and -0.5. The magnitude frequency responses of the subfilters, obtained from two possible parings, are shown in Figure 15.9. The resulting filters are the following:

- proximity pairing: $H_1(z) = (z - 0.5)/(z - 0.9)$ and $H_2(z) = (z + 0.5)/(z + 0.9)$,
- nonproximity pairing: $H_1(z) = (z + 0.5)/(z - 0.9)$ and $H_2(z) = (z - 0.5)/(z + 0.9)$.

The maximum subfilter gain for the proximity pairing design is $G_{max} = 5.0$. For the other pairing, $G_{max} = 15.0$. Given a choice, the proximity paired design would normally be selected.

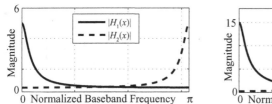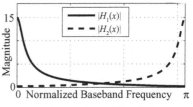

Figure 15.9 Magnitude frequency responses of subfilters. Proximity pairing (left) and nonproximity pairing (right).

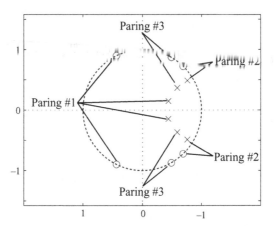

Figure 15.10 Pole-zero distribution of the Chebyshev II filter showing the proximity pairings for the three subfilters.

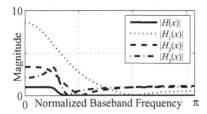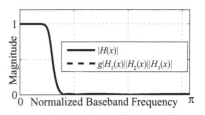

Figure 15.11 System output and individual subsystem responses. The maximum subsystem gain ranges from about 2 to nearly 9. The scaled by "*g*" cascaded response of the subsystems and original transfer function are seen to be in full agreement.

Example: Pole-Zero Paring

Design a sixth-order Chebyshev II low-pass filter that has at least a −50-dB stopband beginning at $f = 0.125f_s$. The filter is to be implemented using a cascade architecture. The pole-zero distribution is displayed in Figure 15.10. The frequency response of each second-order filter is displayed in Figure 15.11. It can be seen that the frequency response of each section is different in shape and maximal value. Other pairs would result in higher localized gain.

MATLAB FIRST-AND SECOND-ORDER SECTIONS

MATLAB contains a collection of programs that recognize the connection between state variables and a cascade filter implementation. It is based on the use of MATLAB's second-order system (sos) format that represents second-order sections. If

$$H_i(z) = \frac{B_i(z)}{A_i(z)} = \frac{b_{0i} + b_{1i}z^{-1} + b_{2i}z^{-n}}{1 + a_{1i}z^{-1} + \ldots + a_{2i}z^{-m}} ; i \in [1, L],$$

then the sos coefficient matrix is given by

$$\text{sos} = \begin{bmatrix} b_{01} & b_{11} & b_{21} & 1 & a_{11} & a_{21} \\ b_{02} & b_{12} & b_{22} & 1 & a_{12} & a_{22} \\ \ldots & \ldots & \ldots & \ldots & \ldots & \ldots \\ b_{0L} & b_{1L} & b_{2L} & 1 & a_{1L} & a_{2L} \end{bmatrix}.$$

The MATLAB cascade filter objects are the following:

tf2sos: The command [sos,g] = tf2sos(b,a) converts a filter having a transfer function $H(z)$ as a collection of L second-order filters and stores the outcome in an sos $L \times 6$ sos matrix. If $H(z)$ is a first-order system, then $b_{2k} = a_{2k} = 0$. The tf2sos function uses a multistep algorithm to determine the second-order section representation by

1. computing poles and zeros of $H(z)$,

2. grouping the zeros and poles into complex conjugate pairs on the basis of proximity, and

3. implementing second-order subfilters.

zp2sos: Program [sos,g] = zp2sos(b,a,k) converts a filter having a transfer function into a collection of L second-order filter sections in the form of an sos $L \times 6$ matrix.

ss2sos: Program [sos,g] = ss2sos(A,B,C,D) converts a filter having a state variable representation into a collection of L second-order filter sections in the form of an sos $L \times 6$ matrix.

The mapping process can be reversed using the following functions.

sos2ss: The command [A,B,C,D] = sos2ss(sos) converts a filter of second-order sections into a filter having an equivalent state-space representation.

sos2tf: The command [b,a] = sos2tf(sos) converts a filter consisting of second-order sections into a transfer function.

Example: Cascade IIRs

Consider the third-order IIR digital having the transfer function

$$H(z) = \frac{N(z)}{D(z)} = \frac{1 - 0.5z^{-1} - 0.315z^{-2} - 0.185z^{-3}}{1 - 0.5z^{-1} + 0.5z^{-2} - 0.25z^{-3}}.$$

Interpreted $H(z)$ as a cascade IIR using $\texttt{tf2sos(N,D)}$ as shown below:

```
N=[1 -.5 -.315 -.0185]; D=[1 -.5 .5 -.25];
[SOS,G]=TF2SOS(N,D)

SOS = 1.0000 -0.8813 0.0000 1.0000 -0.5000 0.0000

       1.0000  0.3813 0.0210 1.0000  0.0000 0.5000

G = 1
```

The outcome is two cascaded sections, namely,

$$H(z) = GH_1(z)H_2(z) = \frac{1 - 0.8813z^{-1}}{1 - 0.5z^{-1}} \frac{1 + 0.3813z^{-1} + 0.021z^{-2}}{1 + 0.5z^{-2}}$$

Once the $Q = 2$ subfilters have been defined, the individual transfer functions can be mapped into a state-determined model using the MATLAB program sos2ss, using the format [A,B,C,D] = sos2ss(SOS,G) for each individual subfilter:

$$A_1 = [0.5]; \quad b_1 = [1]; \quad c_1 = [-0.3813]; \quad d_1 = 1$$

$$A_2 = \begin{bmatrix} 0 & -0.5 \\ 1 & 0 \end{bmatrix}; \quad b_2 = \begin{bmatrix} 1 \\ 0 \end{bmatrix}; \quad c_2 = [0.3813 \quad -0.479]; \quad d_2 = 1.$$

The individual filter sections have their direct II state variable models shown in Figure 15.12.

PARALLEL ARCHITECTURE

Another basic filter structure is the parallel architecture. A parallel collection of Q IIR subfilters can be used to create the transfer function

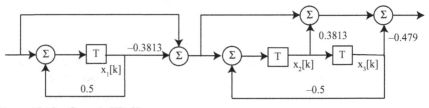

Figure 15.12 Cascade IIR filter.

$$H(z) = K\left(d_0 + \sum_{i=1}^{Q} H_i(z)\right); \quad \sum_{i=1}^{Q} \text{order}(H_i(z)) = N, \tag{15.11}$$

where $H_i(z)$ is a first- or second-order subsystem defined with real coefficients. That is, each first- and second-order system is implemented as a real direct II subfilter. The process of mapping an Nth-order transfer function $H(z)$ into a parallel filter is facilitated by a Heaviside expansion. To illustrate, the Heaviside expansion of a third-order IIR, given by $H(z) = (0.44z^2 + 0.362z + 0.02)/(z^3 + 0.4z^2 + 0.18z - 0.2)$, is

$$H(z) = \frac{\gamma_1 z^{-1}}{1 + p_1 z^{-1}} + \frac{\gamma_2 z^{-1}}{1 + p_2 z^{-1}} + \frac{\gamma_2^* z^{-1}}{1 + p_2^* z^{-1}} = \frac{Az^{-1}}{1 + p_1 z^{-1}} + \frac{Bz^{-2} + Cz^{-1}}{1 + 2\,\text{Re}(p_2) + |p_2|^2\, z^{-2}}.$$

The Heaviside expansion of $H(z)$ is given in terms of

pole location: $p_1 = 0.4000$, $p_2 = -0.4000 - j0.5831$; $p_2^* = 0.4000 + j0.5831$,

Heaviside: $\gamma_1 = 0.24000$, $\gamma_2 = 0.1000 - j0.1458$; $\gamma_2^* = 0.1000 + j0.1458$,

which results in the coefficients A, B, and C, namely,

$A = 0.24$;

$B = 2(\text{Re}(\gamma_2) = 2) \times \text{Re}(0.1 - j0.1458) = 0.20$; and

$C = 2(\text{Re}(\gamma_2) \times \text{Re}(p_2)) + 2(\text{Im}(\gamma_2) \times \text{Im}(p_2)) = 2 \times (0.1) \times (-0.4) + 2 \times (0.5831) \times (-0.1458) = -0.25.$

This produces the parallel architecture shown in Figure 15.13 and modeled by

$$H(z) = \frac{0.24z^{-1}}{1 + 0.4z^{-1}} + \frac{0.2z^{-2} - 0.25z^{-1}}{1 + 0.8z^{-1} + 0.5z^{-2}}.$$

CASCADE/PARALLEL MATLAB SUPPORT

MATLAB's Signal Processing Toolbox contains a collection of objects that can be used to map a transfer function $H(z)$ into a cascade or parallel architecture. These

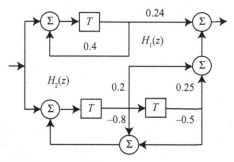

Figure 15.13   Parallel third-order IIR Filter.

operations use the function dfilt to describe a discrete-time filter. The command Hd = dfilt.structure(input1, ...) returns a discrete-time filter having a user-defined structure or architecture. For cascade and parallel architectures, the choices are the following:

dfilt.cascade—filters arranged in series,

dfilt.parallel—filters arranged in parallel,

dfilt.df1sos—direct I collection of second-order sos sections,

dfilt.df1tsos—transpose direct I collection of second-order sos sections,

dfilt.df2sos—direct II collection of second-order sos sections,

dfilt.df2tsos—transpose direct II collection of second-order sos sections.

These programs are based on reducing a given transfer function $H(z)$ into first- or second-order direct I or direct II as standard and transpose architectures all having real coefficients. For sos, an $L \times 6$ matrix of second-order sections, the command dfilt.df1sos creates a cascaded discrete-time direct I filter of second-order sections. The command Hd = dfilt.df1sos(sos) returns a cascade architecture that is built on a collection of second-order direct I filter sections, Hd = dfilt.df1tsos(sos)) for direct I transposed filters, Hd = dfilt.df2sos(sos) for direct II filters, and Hd = dfilt.df2tsos(sos) for direct II transposed filters. The commands tf2sos or zp2sos can be used to create the sos matrix. These provide a plurality of options for implementing cascade filters.Alternatively, the MATLAB command dfilt.cascade(h1, h2,...) returns a discrete-time filter with serial interconnection of two or more filter objects, which are separately designed (see Fig. 15.14) using MATLAB architectural tools (e.g., direct I conversion). The dfilt.parallel(h1, h2,...) command is used to construct a parallel filter (see Fig. 15.14).

Example: MATLAB Cascade and Parallel IIR

Implement a sixth-order, low-pass, elliptic filter as a cascade collection of three direct II second-order filters. This result can be obtained using dfilt.df2sos(s). The magnitude frequency response of the derived filter and the three cascade stages and the three cascade sections are shown in Figure 15.15 using MATLAB's fvtool. The characteristics of each second-order filter are seen to be distinct. It can also be noted that the overall (cascaded) frequency response has a 0 Hz (DC) gain of about

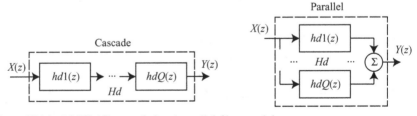

Figure 15.14   MATLAB cascaded and parallel filter module.

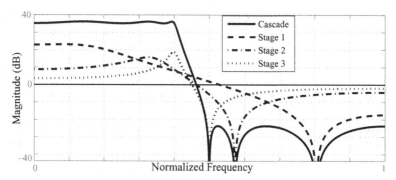

Figure 15.15   Magnitude frequency response of sections $H_1(z)$, $H_2(z)$, $H_3(z)$, and $H(z)$. Display provided by MATLAB's fvtool.

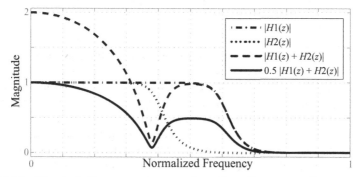

Figure 15.16   Magnitude frequency response of two direct II transpose filters connected in parallel. Notice that $|Hp| \neq |H1| + |H2|$ but $|Hp| = 0.5(|H1| + |H2|)$. Display is provided by MATLAB's fvtool.

36 dB, not 0 dB. The discrepancy can be explained when it is realized that the separate input gain adjustment of $g = 0.0153$ (−36.29 dB), which was not applied to the cascade filter.

To explore a parallel architecture, two eighth-order unit gain low-pass IIR filters $H_1(z)$ and $H_2(z)$ with different passband cutoff frequencies are designed and combined in parallel to form $H(z)$ using dfilt.parallel(Hd1,Hd2). The passband gain of $H(z)$ has a value of 2 which, if cascaded with a memory-less filter with a gain of 0.5, produces a unit gain passband outcome. This is illustrated below and the outcome is displayed in Figure 15.16, using MATLAB's fvtool.

LADDER/LATTICE IIRS

Lattice filters appear in both finite impulse response (FIR) and IIR forms. They have great utility in a number of digital signal processing (DSP) applications, such as speech processing and adaptive filtering. Lattice filters are essentially collections of independent filter sections or stages. Sections can be added or removed without affecting the others. This is unlike other filters where a change in one part of a design can affect the entire design. The order of a lattice filter, for example, can be increased without the need of recomputing the preexisting sections. Lattice filters are also claimed to have a low sensitivity to coefficient round-off error, making them attractive for use with small word length technologies. Another lattice feature is the ability to compute both forward and backward predictions. Forward error predictions (a priori) are used to estimate a current or future value of a signal in noise. Backward error predictions (a posteriori) are used to estimate a past value of a signal in noise.

In general, an Mth-order linear time invariant IIR can be represented as the monic transfer function

$$H(z) = \frac{\sum_{i=0}^{M} b_i z^{-i}}{1 + \sum_{i=1}^{M} a_i z^{-i}} = \frac{B(z)}{A(z)} \tag{15.12}$$

The Mth-order system can be realized as the ladder/lattice IIR filter as shown in Figure 15.17. The filter structure consists of two major subsystems. The first is a lattice section that is similar to the feed-forward structure found in a lattice FIR. This section is defined by a set of reflection or partial correlation (PARCOR) coefficients, with the difference being that the FIR feed-forward structure is replaced by a feedback structure. Attached to feedback taps of the lattice filter is a collection of tap weights that are used to synthesize a unique input–output relationship. One of the known properties of a lattice IIR filter is that the reflection coefficients of a stable lattice filter are bounded by unity (i.e., $|k_i| < 1.0$). The IIR design procedure is similar in form to that developed for lattice FIRs along with a production means for determining the tap-weight coefficient λ_i that defines the ladder section.

Example: IIR Lattice Filter

To motivate the IIR design process, consider the three-stage example having a transfer function $H(z) = (1 + 2z^{-1} + 3z^{-2} + 4z^{-3})/(1 + (21/32)z^{-1} + (21/64)z^{-2} + (1/8)z^{-3})$. The analysis is based on the knowledge that the lattice IIR is defined in terms of the PARCOR coefficients that are computed to be $\{1/2, 1/4, 1/8\}$ as shown in Figure 15.18.

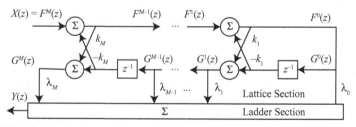

Figure 15.17   Ladder/lattice architecture.

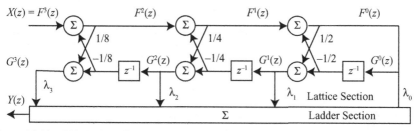

Figure 15.18   Third-order ladder/lattice IIR.

The Signal Processing Toolbox function $\texttt{tf2latc}$ accepts an IIR filter in polynomial form and returns the corresponding reflection coefficients. The process can be reversed using $\texttt{latc2tf}$ to produce an outcome that agrees with the manual lattice filter calculations (see below):

```
num=[1 2 3 4]; den=[1 21/32 21/64 1/8];

[k,v]=tf2latc(num,den);

k=    0.5000  % k₁; 0.2500  % k₂; 0.1250  % k₃;

v =   0.1797  % λ₀; 0.4531  % λ₁; 0.3750  % λ₂; 4.0000  %
λ₃;

[n,d]=latc2tf(k,v);  % reverse process

n =    1    2    3    4

d =    1.0000   0.6563   0.3281   0.1250
```

Recall that MATLAB uses $\texttt{dfilt}$ to describe a discrete-time filter and that $\texttt{Hd} = \texttt{dfilt.structure(input1,...)}$ returns a discrete-time filter (i.e., *Hd*). MATLAB provides a collection of $\texttt{dfilt}$ objects that map a transfer function into a set of lattice architectures. Specifically, they are

| $\texttt{dfilt.latticear}$ | lattice autoregressive (AR) and |
| $\texttt{dfilt.latticearma}$ | lattice autoregressive moving average (ARMA), |

where the AR model refers to an all-pole IIR structure and ARMA has both poles and zeros. The internal structure of an AR lattice filter is shown in Figure 15.17 where the output is taken from point $F^0(z)$ and the coefficients $\lambda_i = 0$. The ARMA filter appears in the form shown in Figure 15.17.

CHAPTER 16

FIXED-POINT EFFECTS

BACKGROUND

Digital filters can be studied in the context of their architecture. Unless this analysis reflects reality, performance prediction can range from being misleading to sometimes completely false. Virtually all digital filter analysis protocols are based on linear models. The veracity of these models can be challenged when nonlinear behavior appears in the form of finite word length effects. However, to ensure successful filter performance under real-world conditions, finite word length effects must be understood, measured, and managed.

FIXED-POINT SYSTEMS

An N-bit fixed-point signed number is represented by one sign bit, I integer bits, and F fractional bits such that $N = I + F + 1$ bits. Such a system is said to have an $[N:F]$ format. The dynamic range of an $[N:F]$ system is bounded by $\pm 2^I$, and the significance of the least significant bit is $Q = 2^{-F}$. Since a fixed-point system possesses both a limited dynamic range and precision, resources must be intelligently managed. If not, the performance can be compromised. The finite word length errors, introduced by a fixed-point system, are

- overflow saturation,
- arithmetic rounding,
- coefficient rounding,
- data scaling, and
- zero-input limit cycling.

Fortunately, corrective actions are generally available and take the form of

- scaling the input or coefficients,
- increasing the system's word length, and
- selecting an alternative architecture or filter model.

Digital Filters: Principles and Applications with MATLAB, First Edition. Fred J. Taylor.
© 2012 by the Institute of Electrical and Electronics Engineers, Inc.
Published 2012 by John Wiley & Sons, Inc.

A high-level design and analysis strategy begins with the translation of a set of filter specifications into a realizable filter's transfer function, and ultimately assigned a specific architecture. The designed filter can be realized as a floating-point or fixed-point filter. The performance of the candidate filter solution will need to be studied using analytical methods or simulations to ensure that the design specifications remain satisfied. If the candidate design fails this test, then the design strategy will need to be altered or modified, and tested until a successful design is obtained, if one exists.

OVERFLOW (SATURATION) EFFECTS

One of the recognized finite word length effects is called run-time register overflow. Of all the error sources, register overflow is potentially the most disastrous. Register overflow can introduce significant nonlinear distortion into a system potentially rendering a filter useless. It is therefore incumbent on the filter designer to eliminate, or at least mitigate, the effects of run-time register overflow. There are some standard precautions that can be used to control this problem in some circumstances. First, if fixed-point arithmetic is used to implement a filter design, it should be 2's complement (2C) because of the modulo(N) property. It is known that an array of N-bit 2C data words can be accumulated without error as long as the final result is known to be a valid N-bit 2C number. The result is correct even if accumulator overflow is encountered during run time. Therefore, bounding a filter's outcome to be a valid N-bit word, the worst case dynamic range requirements of the digital filter, needs to be mathematically and physically certified.

Another method used to suppress the potential effects of register saturation is to use saturating arithmetic. A saturating arithmetic unit, upon detecting an overflow condition, clamps the accumulator output to the most positive or negative value (see Fig. 16.1). That is, the output of an N-bit 2C saturating accumulator, having a maximal dynamic range limit of $\pm 2^{N_0}$, $N_0 \leq N$, is defined to be

$$ACC = \begin{cases} \left(2^{N_0-1}-1\right) & \text{if} \quad ACC \geq 2^{N_0-1} \\ ACC & \text{if} \quad -2^{N_0-1} < ACC < 2^{N_0-1}-1. \\ -2^{N_0-1} & \text{if} \quad ACC \leq -2^{N_0-1} \end{cases} \qquad (16.1)$$

Register overflow can be studied experimentally by continually scaling the input downward until overflows cease to occur. Unfortunately, experimental studies can

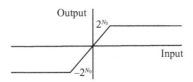

Figure 16.1   Saturating arithmetic unit.

only test a finite number of input sequences. One could potentially eliminate over-flow errors by aggressively scaling the input, but such a policy will also reduce the precision of the resulting filter to a point that the filter is of no practical value. Since dynamic range and precision metrics are inversely related in fixed-point systems, it is important to mathematically define the dynamic range bounds as conservatively as possible. This question is called the binary-point assignment problem.

Example: Run-Time Overflow

Consider an eighth-order low-pass Chebyshev II filter having a transfer function

$$H(z) = 0.0888 \frac{\left(\begin{array}{c} z^8 + 4.43z^7 + 10.76z^6 + 17.46z^5 + 20.48z^4 + 17.46z^3 + \\ 10.76z^2 + 4.43z + 1 \end{array} \right)}{\left(\begin{array}{c} z^8 + 1.10z^7 + 1.97z^6 + 1.55z^5 + 1.22z^4 + 0.61z^3 + 0.24z^2 + \\ 0.061z + 0.008 \end{array} \right)}.$$

The filter is implemented using a [16:15] formatted fixed-point direct II and cascade infinite impulse response (IIR). The system is tested using a unit step function. The simulated step responses obtained from both architectures are shown in Figure 16.2 along with the ideal response. The direct II response is highly distorted due to 93 internal register overflows occurring in the first 101 samples. The cascade filter's response shows far less distortion and suffered only 19 register overflows during the same time period. It can be noted, however, that only the transient response was significantly distorted due to finite word length effect errors. This is due to the nature of the test conducted, namely, a study of the IIR's step response. If the transient response is important, such as motor control applications, the filter responses would significantly differ. In addition, it should be noted that the cascade filter requires 25 multiply-accumulates (MACs) per filter cycle, while the direct II needs only 18 MAC calls per filter cycle. Therefore, the maximum real-time band-width of the direct II can be 40% higher than the cascade. As a result, the fixed-point architecture selection process often raises trade-off issues between bandwidth and precision.

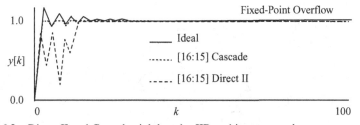

Figure 16.2   Direct II and Cascade eighth-order IIR architectures and responses.

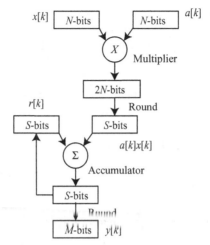

Figure 16.3   Fixed-point MAC unit.

ARITHMETIC ERRORS

The next most severe type of error attributed to finite precision arithmetic effects is called arithmetic error. Consider the fixed-point MAC unit shown in Figure 16.3, which is used to implement SAXPY (S = AX + Y) calls. It is assumed that A and X are N-bit digital words presented to a full-precision multiplier, producing a $2N$-bit full-precision product. The $2N$-bit full-precision product may be reduced to an S-bit word, $S \leq 2N$, before being presented to an S-bit accumulator. The S-bit accumulator is finally reduced to an M-bit final result where $M \leq S$. It is generally assumed that the error associated with the case where $M \ll 2N$ is given by $e = (y[k] - Q_F(y[k]))$ where $Q_F(q)$ denotes the quantization of a number q to a digital word having F fractional bits of precision. The error variance is given by $\sigma^2 = Q^2/12$ where $Q = 2^{-F}$ is the weight of the least significant bit of the M-bit accumulator output having F fractional bits of precision. The problem is not necessarily the modeling the error itself, but determining what happens to the error after it is produced. Since the internally generated errors in an IIR can recirculate and remain within the system indefinitely, monitoring and controlling these errors is essential to successful filter design.

COEFFICIENT SENSITIVITY

A third source of finite word length error is attributed to coefficient rounding. A general model for a monic IIR is

$$H(z) = \frac{\displaystyle\sum_{m=0}^{M} b_m z^{-m}}{1 + \displaystyle\sum_{m=1}^{N} a_m z^{-m}} = d_0 + \frac{\displaystyle\sum_{m=1}^{N} c_m z^{-m}}{1 + \displaystyle\sum_{m=1}^{N} a_m z^{-m}} \qquad (16.2)$$

having only real coefficients. Suppose that a_m, b_m, c_m, d_0 are rounded and replaced by $\hat{a}_m, \hat{b}_m, \hat{c}_m, \hat{d}_0$, the coefficients of $\hat{H}(z)$, where

$$
\begin{aligned}
\hat{a}_m &= Q_V(a_m) = a_m + \Delta a_m, \quad m \in [1, N], \\
\hat{b}_m &= Q_V(b_m) = b_m + \Delta b_m, \quad m \in [1, N], \\
\hat{c}_m &= Q_V(c_m) = c_m + \Delta c_m, \quad m \in [1, N], \\
\hat{d}_0 &= Q_V(d_0) = d_0 + \Delta d_0,
\end{aligned}
\tag{16.3}
$$

and $Q_F(x)$ denotes the quantizing of a coefficient x to F fractional bits of precision. Comparing $H(z)$ into $\hat{H}(z)$, defined in terms of $\hat{a}_m, \hat{b}_m, \hat{c}_m, \hat{d}_0$, is difficult in general due to the nonlinear nature of the problem. Small changes in a denominator coefficient, for example, can radically alter a transfer function's pole location causing, in some cases, instability. Specifically, the poles of the IIR are denoted p_m and are the roots to

$$
D(z) = 1 + \sum_{m=1}^{N} a_m z^{-m} = \prod_{m=1}^{N} \left(1 - p_m z^{-1}\right).
\tag{16.4}
$$

If the coefficient a_m is quantized to values $\hat{a}_m = Q_F(a_m) = a_m + \Delta a_m$, then it follows that

$$
\hat{D}(z) = 1 + \sum_{m=1}^{N} \hat{a}_m z^{-m} = \prod_{m=1}^{N} \left(1 - \hat{p}_m z^{-1}\right),
\tag{16.5}
$$

where $\hat{p}_m = p_m + \Delta p_m, \ m \in [1, N]$. The roots of $\hat{D}(z)$ define the pole locations for a fixed-point filter and are assumed to have moved from their ideal location $z = p_m$ by an amount Δp_m. The incremental change in pole location Δp_m can be modeled:

$$
\Delta p_m = \sum_{k=1}^{N} \frac{p_m}{a_k} \Delta a_k = \sum_{k=1}^{N} \frac{p_i^{N-k} \Delta a}{\prod_{j=1, j \neq i}^{N} (p_i - p_j)}.
\tag{16.6}
$$

Based on a classic partial derivative model, the pole sensitivity is seen to a function of the proximity that the original poles have to each other. It should be apparent, however, that performing this analysis can be very tedious and may have limited value. Furthermore, the partial derivatives used in sensitivity analysis studies are only valid for very small incremental changes in parameters. This makes the value of this type of analysis questionable in the context of a practical fixed-point filter. In lieu of formal mathematical tests, simulation is often used to qualify the effects of coefficient rounding.

Example: IIR Coefficient Sensitivity

A sixth-order Chebyshev I low-pass IIR is designed and implemented as direct II and cascade filters. The filter's transfer function is given by

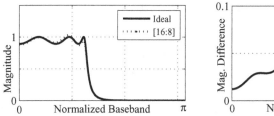 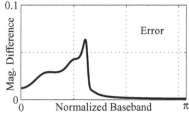

Figure 16.4   Ideal and 8-fractional-bit direct II magnitude frequency response (left) and spectral difference (right).

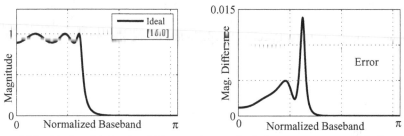

Figure 16.5   Ideal and 8-fractional-bit cascade magnitude frequency responses (left) and spectral difference (right).

$$H(z) = \frac{\begin{pmatrix} 0.0026 + 0.0155z^{-1} + 0.0388z^{-2} + 0.0517z^{-3} + 0.0388z^{-4} + \\ 0.0155z^{-5} + 0.0026z^{-6} \end{pmatrix}}{\begin{matrix} 1 - 2.839z^{-1} + 4.635z^{-2} - 4.806z^{-3} + 3.317z^{-4} - \\ 1.435z^{-5} + 0.3234z^{-6} \end{matrix}}.$$

The filter is implemented as direct II and cascade IIRs with the filter coefficients rounded to a [16:8] format. The analysis of the direct II filter is shown in Figure 16.4.

The analysis of the cascade filter is shown in Figure 16.5. The filter is reduced to three second-order sections having only real coefficients. The coefficients of the second-order section are quantized and the system response is recorded.

The difference between the responses can be attributed to the coefficient rounding. It can be seen that cascade architecture has a slight advantage over direct II but represents a slightly more complex implementation. This observation generally holds when comparing cascade and direct II fixed-point filters.

To illustrate coefficient rounding effects in a state variable format, consider an eighth-order narrow-band band-pass Chebyshev I filter. The filter is to have a 1-dB passband over $f_p \in [0.4, 0.6]f_s$, and 50-dB attenuation in the stopbands of $f_{a1} \in [0.0, 0.2]f_s$ and $f_{a2} \in [0.8, 1.0]f_s$. The filter's transfer function is given by

$$H(z) = \frac{0.00183 - 0.00734z^{-2} + 0.01101z^{-4} - 0.00734z^{-6} + 0.00183z^{-8}}{1 + 3.054z^{-2} + 3.829z^{-4} - 2.292z^{-6} + 0.5507z^{-8}}.$$

The direct II IIR implementation is described in the standard state variable model shown below. The filter is to be implemented with 8-bit and 15-bit fractional coefficient precision. This can be accomplished by rounding the coefficients of the state space model version of the filter to F fractional bits of precision:

$$
A = \begin{bmatrix}
0 & -3.05 & 0 & -3.82 & 0 & -2.29 & 0 & -0.55 \\
1 & 0 & 0 & 0 & 0 & 0 & 0 & 0 \\
0 & 1 & 0 & 0 & 0 & 0 & 0 & 0 \\
0 & 0 & 1 & 0 & 0 & 0 & 0 & 0 \\
0 & 0 & 0 & 1 & 0 & 0 & 0 & 0 \\
0 & 0 & 0 & 0 & 1 & 0 & 0 & 0 \\
0 & 0 & 0 & 0 & 0 & 1 & 0 & 0 \\
0 & 0 & 0 & 0 & 0 & 0 & 1 & 0
\end{bmatrix} ;\quad
b = \begin{bmatrix} 1 \\ 0 \\ 0 \\ 0 \\ 0 \\ 0 \\ 0 \\ 0 \end{bmatrix} ;\quad
c = \begin{bmatrix} 0 \\ -0.013 \\ 0 \\ 0.004 \\ 0 \\ -0.012 \\ 0 \\ 0.0018 \end{bmatrix} ;
$$

$d = 0.008.$

The outcome is shown in Figure 16.6. It can be seen that the 8-fractional-bit design exhibits some passband degradation, while the 15-fractional-bit solution is a good approximation to the ideal.

A common problem encountered in fixed-point filter design is the quantizing of small, but important, filter coefficients to zero. To illustrate, the numerator $N(z)$ of the example IIR is $N(z) = (0.00183 - 0.00734z^{-2} + 0.01101z^{-4} - 0.00734z^{-6} + 0.00183z^{-8})$. The smallest coefficient is 0.00183 and would require at least 9 fractional bits to avoid being quantized to zero. Assigning more fractional bits to the design, while improving the precision of small coefficients, would reduce the maximum dynamic range of the system, exposing it to possible run-time register overflow.

The eighth-order Chebyshev I band-pass IIR studied as a direct II architecture can also be implemented using a cascade architecture. Replicating the analysis performed for a direct II study, the cascade analysis would proceed as follows. Using proximity pairing, the eighth-order Chebyshev I band-pass IIR would be factored into four second-order sections satisfying

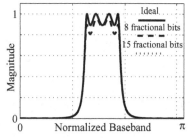 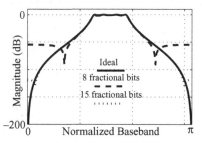

Figure 16.6 Eighth-order band-pass filter showing the magnitude frequency response of the ideal and 8- and 15-fractional-bit direct II filters (left) and in decibels (right).

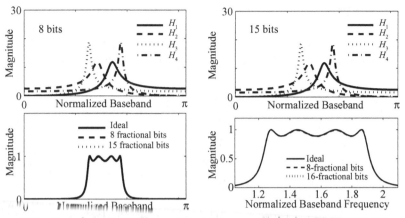

Figure 16.7   Cascade implementation of an eighth order Chebyshev I band-pass IIR using 8 and 15 fractional bits of coefficient precision showing the response of each second-order section. Shown are the ideal filter response plus those having 8 and 15 fractional bits of coefficient precision (upper left and right), and the output response of the cascade filter for each case (lower left) and a zoom expansion of part of the passband (lower right).

$$H(z) = gH_1(z)H_2(z)H_3(z)H_4(z)$$

$$= 0.0018 \frac{1 - 2z^{-1} + z^{-2}}{1 + 0.238z^{-1} + 0.805z^{-2}} \frac{1 - 2z^{-1} + z^{-2}}{1 - 0.238z^{-1} + 0.805z^{-2}}$$

$$\frac{1 - 2z^{-1} + z^{-2}}{1 - 0.585z^{-1} + 0.921z^{-2}} \frac{1 - 2z^{-1} + z^{-2}}{1 + 0.585z^{-1} + 0.921z^{-2}}.$$

The resulting second-order sections are then mapped into a direct II and cascade form. The outcome is shown in Figure 16.7. The fixed-point filters are seen to produce a reasonably good approximation to the ideal magnitude frequency response. This observation is generalizable in that cascade IIRs are normally expected to provide a solution that has a higher coefficient round-off error tolerance.

SECOND-ORDER SECTIONS

The poles of an Nth-order IIR having a transfer function $H(z) = kN(z)/D(z)$ are the roots of $D(z) = 0$. For a direct II architecture, the characteristic equation is denoted $D^{II}(z)$ and is given by Equation 16.7 for the poles located at $z = p_m$. Specifically,

$$D^{II}(z) = D(z) = \sum_{m=0}^{N} a_m^{II} z^{-m} = \prod_{m=0}^{N} \left(1 - p_m^{II} z^{-1}\right). \quad (16.7)$$

Rounding or truncating of the coefficients a_m to F fractional bits of precision, however, would result in a new characteristic polynomial:

$$D_R^{II}(z) = \sum_{m=0}^{N} Q_F(a_m^{II})z^{-m} = \prod_{m=0}^{N}\left(1 - p_m^{II}z^{-1}\right). \tag{16.8}$$

Specifically, $Q_F(a_m^{II}) = a_m + \Delta a_m$ and $p_m^{II} = p_m + \Delta p_m^{II}$. The effect of this pole movement is defined by the collective action of all the quantized feedback coefficients. Suppose that the IIR is implemented with a cascade architecture consisting of a collection of second-order filters. If the coefficients of the cascade architecture are truncated or rounded, then the poles of the fixed-point filter's cascaded characteristic equation is given by

$$D_R^C(z) = \sum_{m=1}^{N/2}(1 + Q_F(a_{1:m}^C)z^{-1} + Q_F(a_{2:m}^C)z^{-2}) = \prod_{m=0}^{N}\left(1 - p_m^C z^{-1}\right), \tag{16.9}$$

where $Q_F(a_{i:m}^C) = a_{i:m} + \Delta a_{i:m}$ and $p_m^C = p_m + \Delta p_m^C$ and, in general, $\Delta p_m^{II} \neq \Delta p_m^C$. The movement of the poles of a cascade architecture are seen to be defined by the rounding or truncation of pairs of feedback coefficients, which are distinctly different than those of a direct II filter. It should therefore be apparent that the pole sensitivity is architecture dependent. Since second-order sections are considered to be digital filter building blocks, they have been the focus of a number of focused studies. If the characteristic equation for the second-order section is

$$D(z) = 1 - 2r\cos(\phi)z^{-1} + r^2 z^{-2}, \tag{16.10}$$

then the second-order poles are located at $z = re^{\pm j\phi}$. A second-order direct II filter section would possess a state matrix A given by

$$A = \begin{bmatrix} 0 & 1 \\ -r^2 & 2r\cos(\phi) \end{bmatrix} \tag{16.11}$$

and that of the quantized second direct II section are given by the determinant of

$$Q_F(A) = \begin{bmatrix} 0 & 1 \\ Q_F(-r^2) & Q_F(2r\cos(\phi)) \end{bmatrix}. \tag{16.12}$$

The eigenvalues (poles) of $Q_F(A)$ are given by

$$\begin{aligned} det(zI - Q_F(A)) &= det\begin{bmatrix} z & -1 \\ -Q_F(-r^2) & z - Q_F(2r\cos(\phi)) \end{bmatrix} \\ &= z^2 - Q_F(2r\cos(\phi))z - Q_F(-r^2) \end{aligned} \tag{16.13}$$

and are displayed in Figure 16.8 for all possible 4-fractional-bit coefficients. It can be seen that the roots of Equation 16.13 provide a nonuniform coverage in the z-plane. The poles lie on a circular loci at various radii from the origin. Notice also that the quantized poles are more dense nearer the periphery of the unit circle and sparse in the interior.

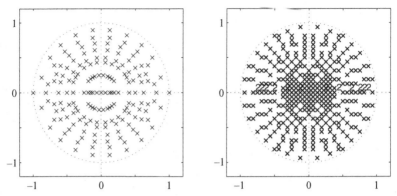

Figure 16.8   Admissible locations of the quantized poles for a signed 4-bit second-order direct II (left) and normal IIR filter (right)

Alternatively, a special second-order architecture, called a normal filter is defined by a state matrix of the form

$$A = \begin{bmatrix} r\cos(\phi) & -r\sin(\phi) \\ r\sin(\phi) & r\cos(\phi) \end{bmatrix}. \tag{16.14}$$

The quantized second-order normal section is given by determinant of the matrix $Q_F(A)$ where

$$Q_F(A) = \begin{bmatrix} Q_F(r\cos(\varphi)) & Q_F(-r\sin(\varphi)) \\ Q_F(r\sin(\varphi)) & Q_F(r\cos(\varphi)) \end{bmatrix}. \tag{16.15}$$

The eigenvalues (poles) of $Q_F(A)$ are given by the roots of

$$det(zI - Q_F(A)) = det \begin{bmatrix} z - Q_F(r\cos(\varphi)) & -Q_F(-r\sin(\varphi)) \\ -Q_F(r\sin(\varphi)) & z - Q_F(r\cos(\varphi)) \end{bmatrix}$$
$$= z^2 - 2Q_F(r\cos(\phi))z + \left(Q_F(r\cos(\varphi))^2 - Q_F(-r\sin(\varphi))^2\right). \tag{16.16}$$

The roots loci of a normal second-order section are displayed in Figure 16.8 for all possible signed 4-fractional-bit coefficients. It can be seen that the direct II poles are nonlinearly distributed in the z-plane and those of a normal filter's poles are more uniformly distributed. Because of this feature, normal architectures are generally assumed to have low coefficient round-off error sensitivity. However, this is conditionally true. If the poles are known to be close to the unit circle, the quantization pattern of the direct II can result in a design having a lower coefficient round-off error budget. However, the analysis should not stop there.

NORMAL IIR

The direct II filter architecture did not have a balanced or symmetric feedback structure. The cascade and parallel architectures were based on first- or second-order direct II sections and therefore also have asymmetric feedback structures. Symmetric feedback is a hallmark of low round-off error sensitivity analog filters, which can be translated into low round-off error normal digital filters. Such digital filters can be expressed as a collection of first- and second-order normal subsystems. In general, a first-order filter has the form

$$H_i(z) = d_{0i} + \frac{c_{i1}z^{-1}}{1 + a_{i1}z^{-1}} = d_{0i} + \frac{c_{i1}z^{-1}}{1 + \lambda_i z^{-1}}, \tag{16.17}$$

where λ_i is the real eigenvalue of the ith subsystem. The one term feedback structure of a first-order system precludes any discussion of feedback symmetry. A second-order section is, however, a different case. Consider the second-order section given by

$$H_i(z) = d_{0i} + \frac{c_{i1}z^{-1} + c_{i2}z^{-2}}{1 + a_{i1}z^{-1} + a_{i2}z^{-2}} = d_{0i} + \frac{N(z)}{\left(1 + \lambda_i z^{-1}\right)\left(1 + \lambda_i^* z^{-1}\right)}, \tag{16.18}$$

where λ_i and its complex-conjugate pair λ_i^* are the complex eigenvalues of the ith subsystem. In particular, let $\lambda_i = \alpha_i + j\beta_i = r_i e^{j\phi_i}$ and note that $1 + a_{i1}z^{-1} + a_{i2}z^{-2} = (1 - (\alpha_i + j\beta_i))(1 - (\alpha_i - j\beta_i))$. It then follows that $a_{i1} = -2\alpha$ and $a_{i2} = \alpha^2 + \beta^2$. The direct II model immediately follows, and it is given by the 4-tuple:

$$A_i = \begin{bmatrix} 0 & 1 \\ -(\alpha_i^2 + \beta_i^2) & 2\alpha_i \end{bmatrix}; \quad b_i = \begin{bmatrix} 0 \\ 1 \end{bmatrix}; \quad c_i = \begin{bmatrix} c_{i1} \\ c_{i2} \end{bmatrix}; \quad d_i = d_{i0}. \tag{16.19}$$

A linear transformation of $[A_i, b_i, c_i, d_i]$ can be considered using the transform matrix T given by

$$T_i = \begin{bmatrix} -\beta_i & 0 \\ -\alpha_i & 1 \end{bmatrix}; \quad T_i^{-1} = \begin{bmatrix} -1/\beta_i & 0 \\ -\alpha_i/\beta_i & 1 \end{bmatrix}. \tag{16.20}$$

The similarity transform T has the ability to create a new architecture, called a normal architecture, having a state 4-tuple representation:

$$\hat{A}_i = T_i A_i T_i^{-1} = \begin{bmatrix} \alpha_i & -\beta_i \\ \beta_i & \alpha_i \end{bmatrix},$$

$$\hat{b}_i = T_i b_i = \begin{bmatrix} 0 \\ 1 \end{bmatrix},$$

$$\hat{c}_i^T = c_i^T T_i = [-(c_2 + c_1\alpha)/\beta \quad c_1], \tag{16.21}$$

$$\hat{d}_i = d_i,$$

which is interpreted in Figure 16.9.

It can be seen that the feedback structure is now symmetric. A mathematical property possessed by a normal architecture is $\hat{A}_i \hat{A}_i^T = \hat{A}_i^T \hat{A}_i$, which produces a diagonal matrix and a determinant $\alpha^2 + \beta^2$. A more generalized embodiment of

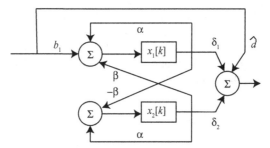

Figure 16.9 Normal filter structure derived from a direct II specification.

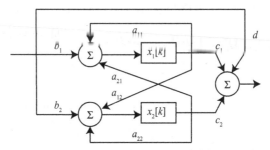

Figure 16.10 Normal filter structure defined by a normal [A, b, c, d] state description.

symmetric feedback architecture is shown in Figure 16.10. The input–output transfer function of the displayed system can be computed to be

$$H(z) = d_0 + \frac{\left((b_1 c_1 + b_2 c_2) z^{-1} + (b_1 a_{12} c_2 + b_2 a_{21} c_1 - b_1 a_{22} c_1 - b_2 a_{11} c_2 +) z^{-2} \right)}{1 - (a_{11} + a_{22}) z^{-1} + (a_{11} a_{22} - a_{12} a_{21}) z^{-2}}. \qquad (16.22)$$

One means of developing a mathematical connection between an arbitrary second-order system and the normal architecture is to use a change of basis, or linear transform of the form

$$T = \begin{bmatrix} \cos(\phi) & -\sin(\phi) \\ \sin(\phi) & \cos(\phi) \end{bmatrix}. \qquad (16.23)$$

The matrix T is called a rotational matrix for obvious reasons. Applying T to the normal architecture shown in Figure 16.10 results in $\hat{A} = TAT^{-1} = A; \hat{b} = TB; \hat{c}^T = c^T T^{-1}; \hat{d} = d$, which maintains the form of A and d, but alters b and c. The mapping T is nonunique, but can be reduced to an algebraic formula. This is accomplished by using a Heaviside expansion of $H(z)$ to form

$$H(z) = d_0 + \sum_{i=1}^{N} \frac{n_i z^{-1}}{1 - \lambda_i z^{-1}}, \qquad (16.24)$$

where n_i is the ith Heaviside coefficient satisfying

$$n_i = \frac{(z - \lambda_i)}{z} H(z) \Big|_{z=\lambda_i}.$$ (16.25)

It is assumed that complex poles come in complex-conjugate pairs. Combining two such complex first-order sections results in

$$H(z) = \cdots + \frac{n_i}{\left(1 - \lambda_i z^{-1}\right)} + \frac{n_i^*}{\left(1 - \lambda_i^* z^{-1}\right)} + \cdots,$$ (16.26)

where n_i and n_i^* are complex-conjugate pairs and $\lambda_i = r_i e^{j\phi_i}$ with $\lambda_i^* = r_i e^{-j\phi_i}$. The general normal architecture is defined in terms of these parameters. Specifically, define A_i to be

$$A_i = r_i \begin{bmatrix} \cos(\phi_i) & -\sin(\phi_i) \\ \sin(\phi_i) & \cos(\phi_i) \end{bmatrix}.$$ (16.27)

Due to the nonuniqueness of the change of architecture mapping, one can require the elements of the vectors b_i and c_i to satisfy

$$b_i^T = (b_{i,1} \quad b_{i,2}),$$
$$b_{i,1}^2 + b_{i,2}^2 = 2(1 - r_i^2),$$
$$b_{i,2} = \left(\frac{r_i^2 \sin(2\phi_i) + \sqrt{1 + r_i^4 - 2r_i^2 \cos(2\phi_i)}}{1 - r_i^2 \cos(2\phi_i)} \right) b_{i,1},$$
$$c_i^T = (c_{i,1} \quad c_{i,2}),$$ (16.28)
$$c_{i,1} = \frac{\left(b_{i,1}(n_i + n_i^*) + b_{i,2}(jn_i - jn_i^*)\right)}{1 - r_i^2 \cos(2\phi_i)},$$
$$c_{i,2} = \frac{\left(b_{i,1}(jn_i^* - jn_i) + b_{i,2}(n_i + n_i^*)\right)}{1 - r_i^2 \cos(2\phi_i)}.$$

Example: Normal IIR

Consider a filter having a transfer function

$$H(z) = \frac{z^2}{z^2 - 2r\cos(\phi)z + r^2} = \frac{z^2}{z^2 - z + 1/2} = 1.0 + \frac{0.5}{\left(z - e^{j\pi/4}/\sqrt{2}\right)} + \frac{0.5}{\left(z - e^{-j\pi/4}/\sqrt{2}\right)}.$$

The eigenvalues are $\lambda_1 = 0.707 e^{j\pi/4}$ and $\lambda_1^* = 0.707 e^{-j\pi/4}$ having residues (Heaviside coefficients) $n_i = n_i^* = 0.5$. It then follows from the recipe provided that

$$A_i = r_i \begin{bmatrix} \cos(\phi_i) & -\sin(\phi_i) \\ \sin(\phi_i) & \cos(\phi_i) \end{bmatrix} = \begin{bmatrix} 0.5 & -0.5 \\ 0.5 & 0.5 \end{bmatrix},$$

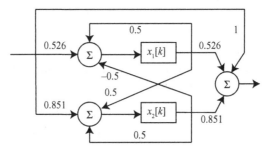

Figure 16.11 Normal filter architecture.

$$b_i^{\mathrm{T}} = (b_{i,1} \quad b_{i,2}),$$

$$h_{i,1}^2 + h_{i,2}^2 = 2\left(1 - r_i^2\right) = 2\left(1 - 1/2\right) = 1.0,$$

$$b_{i,2} = \left(\frac{r_i^2 \sin(2\phi_i) + \sqrt{1 + r_i^4 - 2r_i^2 \cos(2\phi_i)}}{1 - r_i^2 \cos(2\phi_i)}\right) b_{i,1},$$

$$= \left(\frac{(1/2)\sin(\pi/2) + \sqrt{1 + 1/4 - \cos(\pi/2)}}{1 - (1/2)\cos(\pi/2)}\right) b_{i,1} = \frac{\left(1 + \sqrt{5}\right)}{2} b_{i,1} = 1.618 b_{i,1}.$$

Substituting, one obtains $b_{i,1}^2 + (1.618)^2 b_{i,1}^2 = 3.618 b_{i,1}^2 = 1.0$, $b_{i,1} = 0.526$, $b_{i,2} = 0.851$.

Completing the c coefficients, and noting that $\left(n_i + n_i^*\right) = 1$; $\left(jn_i - jn_i^*\right) = 0$, it follows that

$$c_{i,1} = \frac{\left(b_{i,1}\left(n_i + n_i^*\right) + b_{i,2}\left(jn_i - jn_i^*\right)\right)}{\left(1 - r_i^2 \cos(2\phi_i)\right)} = \frac{b_{i,1}(1) + b_{i,2}(0)}{\left(1 - (1/2)\cos(\pi/2)\right)} = \frac{b_{i,1}(1) + b_{i,2}(0)}{1} = b_{i,1},$$

$$c_{i,2} = \frac{\left(b_{i,1}\left(jn_i^* - jn_i\right) + b_{i,2}\left(n_i + n_i^*\right)\right)}{\left(1 - r_i^2 \cos(2\phi_i)\right)} = \frac{b_{i,1}(0) + b_{i,2}(1)}{\left(1 - (1/2)\cos(\pi/2)\right)} = \frac{b_{i,1}(0) + b_{i,2}(1)}{1} = b_{i,2}.$$

Assembling the parts shown in Figure 16.11 for the ith subsystem, one obtains

$$H(z) = 1 + \frac{\left(\left(b_1^2 + b_2^2\right)z^{-1} + \left(-0.5b_1^2 - 0.5b_2^2 - 0.5b_1b_2 + 0.5b_1b_2\right)z^{-2}\right)}{1 - \left(a_{11} + a_{22}\right)z^{-1} + \left(a_{11}a_{22} - a_{12}a_{21}\right)z^{-2}}$$

$$= 1 + \frac{\left((1)z^{-1} + (-0.5)z^{-2}\right)}{1 - (1)z^{-1} + (0.5)z^{-2}}.$$

SCALING

Register overflow can be inhibited by reducing the dynamic range of the input. Scaling the input by a positive constant $K < 1$ reduces the chance of run-time overflow. Unfortunately, scaling also reduces the output precision. In particular, the precision of a scaled data word will be reduced by k-bits, where $k = \log_2(K)$. Equivalently, the output precision will be reduced by a like amount. Therefore, the error variance, compared with the unscaled system, can be expressed as

$$\sigma_K^2 = \frac{(KQ)^2}{12} = K^2\sigma^2. \qquad (16.29)$$

As a result, scaling (if required) should only be used minimally to the point where overflow is controlled. Expanding the word length of a system from N-bits to $(N + M)$-bits will increase the dynamic range of the system by M-bits. This is done, however, at an additional system cost and possible reduction in the maximum real-time bandwidth. In the context of a modern digital signal processing (DSP) microprocessor, the logical choices are 16 bits or 24 bits.

LIMIT CYCLING

Another finite word length effect is called zero input limit cycling, or simply limit cycling. Limit cycling causes a digital filter to produce small amplitude changes to appear at the system's output during periods when the input is zero. In a voice communication application, limit cycling can manifest itself as an undesirable "clicking" sound that is audible during quiet (unvoiced) periods. The first-generation DSP microprocessors were imprecise (e.g., 8 bits), and as a result, limit cycling was an annoying problem that could be reduced through serious engineering labor. With the advent of 16-, 24-bit, and floating-point processors, limit cycling has become a secondary problem. Limit cycling, it can be noted, is caused when the response of an unforced stable system does not successfully decay to zero due to finite word length effects. Consider the simple first-order system $y[k] = ay[k - 1] + x[k]$. If $|a| < 1$, then ideally $y[k] \rightarrow 0$ when $x[k] = 0$. If the filter is implemented in fixed point, then $y[k] = Q_F[ay[k - 1]] + x[k]$ where $Q_F[q[k]]$ denotes the quantized value of $q[k]$ to F fractional bits of precision. Refer to Figure 16.12 and suppose that at sample instance k, the first-order system has a value $y[k]$ and an input $x[k] = 0$ for all $k > K$. If the decay rate of the quantized system is too slow to allow the output to decay by an amount less than half a quantization interval (i.e., $\Delta[k] = (y[k] - y[k - 1]) < Q/2$), then $y[k]$ would be rounded back to its previous value $y[k - 1]$. As such, the output would never be able to decay to zero and would have some constant off-set (possibly oscillating), which is called limit cycling.

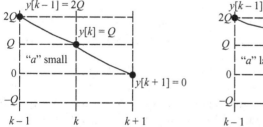 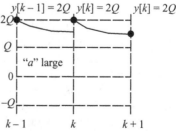

Figure 16.12   Limit cycling interpretation for $y[k + 1] = ay[k]$. Response for small value of "a" is rapid and converges to zero (left). Response for large value of "a" is slow and does not converge to zero (right).

TABLE 16.1. First-Order System Limit Cycling Example

k	$a = 1/2 = 0 \lozenge 1000$	$a = -1/2 = 1 \lozenge 1000$	$a = 3/4 = 0 \lozenge 1100$	$a = -3/4 = 1 \lozenge 1100$
0	0 $\lozenge$ 1111 (15/16)	0 $\lozenge$ 1111 (15/16)	0 $\lozenge$ 1111 (15/16)	1 $\lozenge$ 1111 (15/16)
1	0 $\lozenge$ 1000 (8/16)	1 $\lozenge$ 1000 (−8/16)	0 $\lozenge$ 1011 (11/16)	1 $\lozenge$ 0101 (−11/16)
2	0 $\lozenge$ 0100 (4/16)	0 $\lozenge$ 0100 (4/16)	0 $\lozenge$ 0110 (6/16)	0 $\lozenge$ 0110 (6/16)
3	0 $\lozenge$ 0010 (2/16)	1 $\lozenge$ 1110 (−2/16)	0 $\lozenge$ 0101 (5/16)	1 $\lozenge$ 1100 (−4/16)
4	0 $\lozenge$ 0001 (1/16)	0 $\lozenge$ 0001 (1/16)	0 $\lozenge$ 0100 (4/16)	0 $\lozenge$ 0011 (3/16)
5	0 $\lozenge$ 0001 (1/16)	1 $\lozenge$ 1111 (−1/16)	0 $\lozenge$ 0011 (3/16)	1 $\lozenge$ 1101 (−3/16)
6	0 $\lozenge$ 0001 (1/16)	0 $\lozenge$ 0000 (0/16)	0 $\lozenge$ 0010 (2/16)	0 $\lozenge$ 0010 (2/16)
7	0 $\lozenge$ 0001 (1/16)	0 $\lozenge$ 0000 (0/16)	0 $\lozenge$ 0010 (2/16)	1 $\lozenge$ 1111 (−1/16)
8	0 $\lozenge$ 0001 (1/16)	0 $\lozenge$ 0000 (0/16)	0 $\lozenge$ 0010 (2/16)	0 $\lozenge$ 0001 (1/16)
9	0 $\lozenge$ 0001 (1/16)	0 $\lozenge$ 0000 (0/16)	0 $\lozenge$ 0010 (2/16)	1 $\lozenge$ 1111 (−1/16)
k	Limit cycling at $y[k] = 1/16$	No limit cycling	Limit cycling at $y[k] = 2/16$	Limit cycling at $y[k] = (1/16)(-1)^k$

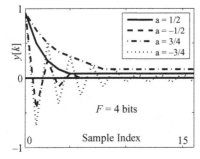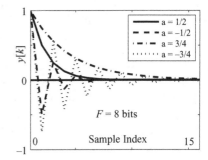

Figure 16.13   Limit cycling for the unforced system $y[k] = ay[k-1] + x[k]$, $x[k] = 0$, using a 4- and 8-bit rounded integer arithmetic.

Example: Limit Cycling

Suppose a system $y[k] = ay[k-1] + x[k]$ has an [N:4] format and $a = 1/2, -1/2, 3/4$, and $-3/4$. Investigate the zero input limit cycling properties of the system. The implementation of these four filters would yield the results in the data as interpreted in Table 16.1 for a 5-bit 2C implementation using an initial condition $x[0] = 1 \rightarrow 0 \lozenge 1111$, where $\lozenge$ denotes the binary point. It can be seen that zero input limit cycling occurs in certain instances.

The 4 zero input responses are simulated for $F = 4$- and 8-bit solutions in Figure 16.13. The filter is initialized to $x[0] = 1.0$ and run. It can be seen that limit cycling can take place under certain circumstances, and that the severity of limit cycling is a function of decay rate of the response (i.e., "a") and the number of bits of arithmetic precision maintained after multiplication. For contemporary 16- or 24-bit designs, especially those using full-precision multipliers and extended precision accumulators, the effect of limit cycling is generally very small or negligible.

CHAPTER 17

IIR ARCHITECTURE ANALYSIS

OVERFLOW PREVENTION

The most grievous finite word length effect is called register overflow. Register overflow can occur whenever the dynamic range requirements of a system variable or parameter exceed the dynamic range capabilities of the filter. Register overflow can introduce major run-time nonlinear distortions into a system's output that can render a filter unusable. In the most severe case, the output may suddenly go from a large positive value to a large negative value. This obviously can create great havoc in audio, servomechanism, biomedical, or other applications. Therefore, register overflow is a problem that must be controlled using detailed mathematical analysis or the intelligent use of computer simulations. A naïve approach to overcoming the overflow obstacle, also called the binary-point assignment problem, is to simply scale a system's input gain so that signal levels do not exceed a filter's specified dynamic range limits. However, scaling will also reduce the filter's precision (data quality), possibly to an unacceptable level. Therefore, a more formal mathematical framework is required.

L_p NORM BOUNDS

A single-input single-output Nth-order proper at-rest infinite impulse response (IIR) digital filter can be reduced to a state variable model (see Fig. 17.1):

$$\begin{aligned} x[k+1] &= Ax[k] + bu[k] \ \{\text{state equation}\}, \\ y[k] &= c^T x[k] + du[k] \ \{\text{output equation}\}. \end{aligned} \tag{17.1}$$

An advantage of a state variable architectural model is found in the fact that state variables directly correlate to the information that resides within the filter's shift registers. Finite word length effect errors produced within an IIR will eventually migrate to the shift register level where they can be observed, quantified, and potentially controlled. This is the key to solving the binary-assignment problem. Fortunately, classical analysis methods have been developed that can be used to compute a state bound in an L_p sense, where the L_p norm of a causal time series $s[k]$ is denoted $\|s\|_p$ and is given by

Digital Filters: Principles and Applications with MATLAB, First Edition. Fred J. Taylor.
© 2012 by the Institute of Electrical and Electronics Engineers, Inc.
Published 2012 by John Wiley & Sons, Inc.

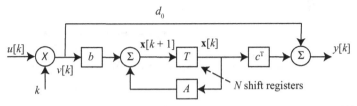

Figure 17.1 State variable model of an IIR showing the states residing at the shift register level.

$$\|s\|_p = \left(\sum_{k=0}^{\infty} |s[k]|^p\right)^{1/p} ; \quad p > 0. \tag{17.2}$$

In the case where $p \to \infty$, $\|s\|_\infty = \max(|s[k]|)$ for all $k \geq 0$.

L_2 OVERFLOW PREVENTION

Developing a state bound begins with computing the impulse response measured at a shift register location. If the impulse response is measured at the ith state, denoted $h_i[k]$, then the forced response at this location is given by the linear convolution sum $y_i[k] = h_i[k] * v[k]$. From Holder's inequality,* it follows that

$$|y_i[k]| \leq \sum_{m=0}^{\infty} |h_i[k-m]v[m]| \leq \|h_i\|_p \|v\|_q; \quad \frac{1}{p} + \frac{1}{q} = 1. \tag{17.3}$$

Observe that if $\|v\|_q \leq 1$, then the state bound at the ith register location will be bounded by $\|h_i\|_p$. This can serve as a measure of the dynamic range requirement for the ith state. A common assumption is that $p = q = 2$ (i.e., L_2 norms). Then, $y_i[k]$ will be bounded if both $h_i[k]$ and $v[k]$ are bounded in an L_2 sense. It may also be recalled that a signal having a finite L_2 norm is of finite energy. Classifying signals on the basis of finite energy (not to be confused with power) is a natural way to partition the signal space. Methods of possessing signals of finite energy have been extensively studied and cataloged by engineers leaving in its wake a rich mathematical legacy. Another advantage of working with L_2 norms is that this study is well developed in the context of signals and linear systems in the time and frequency domains. Recall that the forced or inhomogeneous state response of a state-determined system to an arbitrary input $v[k]$ is given by the convolution sum

$$x[k] = \sum_{n=0}^{\infty} A^k bv[k+1-n]. \tag{17.4}$$

* Holder's inequality states that $\Sigma|x_i y_i| \leq \|x\|_p \|y\|_q$ such that $1/p + 1/q = 1$.

Assume, for illustrative purposes, that the input $v[k]$ is a white zero mean, unit variance wide sense stationary random process. The whiteness of $v[k]$ implies that the signal's autocorrelation function satisfies $E(v[k - n]v[k - m]) = \delta[n - m]$. This leads to the definition of a non-negative definite $N \times N$ matrix K, where

$$K = E(x[k]x^{\mathrm{T}}[k]) = \sum_{n=0}^{\infty} \sum_{m=0}^{\infty} A^n bE(v[k+1-n]v[k+1-m])b^{\mathrm{T}}(A^{\mathrm{T}})^m = \sum_{n=0}^{\infty} A^n bb^{\mathrm{T}}(A^{\mathrm{T}})^n.$$

(17.5)

The energy measured at the ith state location corresponds to the K_{ii} on-diagonal value of K as shown below:

$$K_{ii} = \sum_{k=0}^{\infty} x_i^2[k] = \|x_i\|_2^2.$$

(17.6)

That is, the ith on-diagonal element of K is the L_2 norm squared value of $x_i[k]$ for the case where the input signal is a white random process having unit variance. Thus, Equation 17.6 can provide a gateway to determining state norms provided, of course, that the elements of the matrix K can be efficiently computed. Fortunately, such methods exist.

L_2 NORM DETERMINATION

Suppose, once again, that the input process is a white zero mean wide sense stationary random process. Working with the next state value (i.e., $x[k + 1] = Ax[k] + bv[k]$) and Equation 17.5, the following results:

$$E(x[k+1]x^{\mathrm{T}}[k+1]) = AE(x[k]x^{\mathrm{T}}[k])A^{\mathrm{T}} + bE(v[k]^2)b^{\mathrm{T}} + AE(x[k]v[k])b^{\mathrm{T}}$$
$$+ bE(v[k]\vec{x}^{\mathrm{T}}[k])A^{\mathrm{T}},$$

(17.7)

which simplifies once the a priori knowledge that $E(v[k]) = 0$, $E(x[k]v[k]) = 0$, and $E(v[k]^2) = 1$ is substituted into Equation 17.7. Specifically,

$$E(x[k+1]x^{\mathrm{T}}[k+1]) = E(x[k]x^{\mathrm{T}}[k]) \hat{=} K = AKA^{\mathrm{T}} + bb^{\mathrm{T}}.$$

(17.8)

The $N \times N$ matrix K is often referred to as a Lyapunov stability matrix and is extensively used in the study of state-determined system stability. If K can be readily computed, then the on-diagonal terms can then be used to establish L_2 norm squared state bound on the states, namely, $K_{ii} = \|x_i[k]\|_2^2$. If the system $[A, b, c, d]$ is stable, then the elements of K are bounded. The value of K, from Equation 17.8, can be computed by using the MATLAB procedure shown below:

$$Q = bb^{\mathrm{T}}; K = \mathrm{dlyap}(Q, A) \% \text{ MATLAB Lyapunov function.}$$

(17.9)

The on-diagonal terms of K provide a direct measure of the L_2 norm squared bounds on the state vectors. Once the analysis is performed for a given system

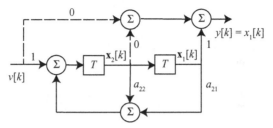

Figure 17.2   State reassignment to isolate a specific state variable shown as $y[k] = x_1[k]$.

$[A, b, c, d]$, the maximum K_{ii} can be determined such that $K_{max} = \text{maximum}(K_{ii})$. To protect the states from encountering run-time dynamic range overflow, the data format $[N . F]$ must be carefully chosen. In the context of K_{max}, the size of the integer field I in bits (i.e., $I = N - F - 1$) should be

$$I = \lceil \log_2 \sqrt{K_{max}} \rceil \tag{17.10}$$

in bits. That is, at least I integer bits need to be allocated to the N-bit data field in order to protect against run-time overflow in an L_2 sense.

The L_2 norm squared bound of a simple second-order IIR can be experimentally determined by driving the system with an input signal having a unit L_2 norm squared, and measuring the L_2 norm or norm squared state bound at the state level. For a direct II architecture, states $x_1[k]$ and $x_2[k]$ are serially (directly) connected. Therefore, the state bound on $x_1[k]$ is the same as the bound on $x_2[k]$. The state $x_1[k]$ can be isolated by assigning the output $y[k]$ to be exclusively $y[k] = x_1[k]$. This can be accomplished by setting $c^T = (1, 0)$ and $d = 0$ to extract $x_1[k]$. This is graphically interpreted for the isolation of state $x_1[k]$ of a second-order direct II section in Figure 17.2. This mapping convention can be used to map selected states of an arbitrary system to the output by modifying the state 4-tuple $[A, b, c, d]$ as follows:

1. Retain A and b.
2. Replace d with $d = 0$.
3. Replace existing c vector with $c = \{\dots, 1 \ (i\text{th location}), \dots\}$ to complete the assignment $y[k] = x_i[k]$.

Isolating $x_2[k]$ would logically require that $c^T = (0, 1)$ and $d = 0$, resulting in $y[k] = x_2[k]$.

Example: L_2 State Bound

Consider the second-order system having a transfer function

$$H(z) = \frac{z - 0.5}{z^2 - z + 0.5} = \frac{z^{-1} - 0.5z^{-2}}{1 - z^{-1} + 0.5z^{-2}}$$

implemented using a traditional direct II architecture having a state model $[A, b, c, d]$ where

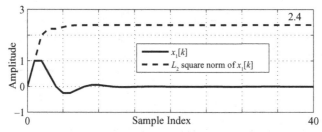

Figure 17.3 L_2 norm squared and state $x_1[k]$ impulse response trajectories.

$$A = \begin{pmatrix} 0 & 1 \\ -0.5 & 1 \end{pmatrix}; \quad b = \begin{pmatrix} 0 \\ 1 \end{pmatrix}; \quad c = \begin{pmatrix} 1 \\ -0.5 \end{pmatrix}; \quad d = 0.$$

From $[A, b, c, d]$, the filter can be explored analytically or experimentally. The resulting Lyapunov matrix K, after Equation 17.9, is:

$$K = \begin{bmatrix} 2.4 & 1.6 \\ 1.6 & 2.4 \end{bmatrix}.$$

State $x_1[k]$ for example, can be assigned to the output port by setting $c_2 = d = 0$ and $c_1 = 1$, resulting in $y[k] = x_1[k]$ as shown in Figure 17.2. If the input $v[k]$ is chosen to be a unit impulse, having a unit L_2 norm, then the resulting output achieves a value $|x_1[k]|_2^2 = |y[k]|_2^2 \leq 12/5 = 2.4$ as shown in Figure 17.3. This is consistent with the MATLAB-computed Lyapunov matrix.

L_2 NORM CAVEAT

The L_2 norm is not strong enough to serve as a reliable method of eliminating run-time register overflow in practice. Most condemning is that its use is restricted to inputs having a bounded L_2 norm. Unfortunately, virtually any interesting or physically meaningful signal (impulse is an exception) has an infinite L_2 norm. For example, for $s[k] = \cos(\omega_0 k)u[k]$, a causal cosine wave, $\|s[k]\|_2 = \infty$. As such, the L_2 norm analysis process is flawed since it does not recognize real-world signals.

Example: Experimental L_2 Norm

Analyze a unit gain low-pass second-order Butterworth IIR having a state variable model

$$A = \begin{bmatrix} 0.2779 & -0.4152 \\ 0.4152 & 0.8651 \end{bmatrix}; \quad b = \begin{bmatrix} 0.5872 \\ 0.1908 \end{bmatrix}; \quad c = \begin{bmatrix} 0.1468 \\ 0.6594 \end{bmatrix}; \quad d = 0.0675.$$

The output impulse response $y[k]$, plus states $x_1[k]$ and $x_2[k]$, are shown in Figure 17.4, converging to zero in about 18 samples. Using MATLAB's `norm` command,

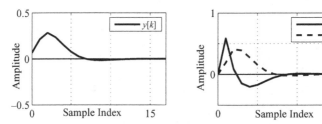

Figure 17.4   System's input–output impulse response (left), and input state impulse response measured at state locations $x_1[k]$ and $x_2[k]$ (right).

the l_∞ norm can be calculated. The L_2^2 squared output norms are computed to be 0.2142, and the l_2^2 state norms are both 0.4592 (using MATLAB), even though they are derived from distinctly different trajectories.

L_∞ NORM BOUNDS

In the pantheon of norms, the L_2 norm is the most popular. In the context of Holder's inequality, an L_2-bounded signal ($p = 2$) convolved with another L_2-bounded system ($q = 2$) impulse response results in a bounded outcome. Unfortunately, except for an impulse, no meaningful signal has a finite L_2 norm. Therefore, the L_2 method is of little practical value. A potentially superior measure is the L_∞ norm where, $p = \infty$ requires $q = 1$ (i.e., $1/p + 1/q = 1$). If the input signal $v[k]$ is bounded in the time domain by unity in an L_1 sense (i.e., $\|v[k]\|_1 = 1$), then $|x_i[k]|$ is bounded by $v[k]\|_1\|h_i[k]\|_\infty$, where $h_i[k]$ is impulse response measured at the output of the ith shift register location (i.e., $x_i[k]$). Unfortunately, many interesting and useful signals $v[k]$ also have an unbounded L_1 norm. However, a different case can be made in the frequency domain. Suppose $V[n]$ is the discrete Fourier transform (DFT) of $v[k]$, $H_i[n]$ is the DFT of $h_i[k]$, and $X_i[k]$ is the DFT of $x_i[k]$. Then theory states that $|X_i[n]|$ is bounded by $\|H_i[n]\|_\infty\|V[n]\|_1$, which simplifies to $|X_i[n]| \leq \|H_i[n]\|_\infty$ if $\|V[n]\|_1 \leq 1$. The use of the L_∞ norm in the frequency domain can be motivated as follows. Let the discrete-time Fourier transform (DTFT) of $h_i[k]$, the impulse response measured at the ith state location, be given by $H_i(e^{j\phi})$ over the baseband range $-\pi \leq \phi \leq \pi$. The L_∞ frequency-domain norm $\|H_i[n]\|_\infty = \text{maximum}(|H_i(e^{j\phi})|) = H_i^{max}(e^{j\phi})$, the maximum magnitude frequency response value across the baseband. Based on Holder inequality, consider an input signal $v[k]$ having an L_1 frequency-domain norm $\|V(e^{j\phi})\|_1 = V$. Upon passing $v[k]$ through a linear filter having an L_∞ frequency-domain norm of $\|H_i(e^{j\phi})\|_\infty = H_i^{max}$, then the output measured at the ith state location will then be bounded by VH_i^{max}. Furthermore, suppose that the inputs to the filter are restricted to be a monotone sinusoid having a unit amplitude with phase $\omega t = \phi_0$. Using Euler's equation, it follows that $v(e^{j\phi}) = 0.5\delta(\phi - \phi_0) \pm 0.5\delta(\phi + \phi_0)$ and

$$\left\|V\left(e^{j\phi}\right)\right\|_1 = \frac{1}{2}\delta(\phi - \phi_0) + \frac{1}{2}\delta(\phi + \phi_0) = \frac{1}{2}(1+1) = 1. \tag{17.11}$$

That is, the L_1 norm of the unit amplitude sinusoid $V(e^{j\phi})$ in the frequency domain is unity. Based on this assumption, the output measured at the ith state locations is bounded by

$$\left| X_i(e^{j\phi}) \right| \leq \left\| V\left(e^{j\phi}\right) \right\|_1 \left\| H_i\left(e^{j\phi}\right) \right\|_\infty = \left\| H_i\left(e^{j\phi}\right) \right\|_\infty \leq H_i^{\max}. \tag{17.12}$$

This states that for a unit amplitude sinusoidal input, the maximal steady-state output at the ith state location is bounded by $H_i^{\max}$ in the frequency domain. This result is intuitively valid and provides motivation for the use of an L_∞ norm in the frequency domain. The L_∞ state bound, unfortunately, assumes that the worst case input is a sinusoid of the form $v[k] = \cos(\omega_0 k T_s)$. This, in turn, presumes that the filter will be evaluated at steady state, and that any overflows that may have occurred took place during the transient period and have since left the system. This obviously still leaves the design exposed to possible overflow errors if the true worst case input is not a pure sinusoid. However, the L_∞ bound is easy to compute using a DFT or fast Fourier transform (FFT). Protecting the system for register overflow in an L_∞ sense would require the data have an integer field of I-bits where $I = \lceil \log_2(\text{maximum}(H_i^{\max})) \rceil$. However, it should be recalled that this theory is only valid for the steady-state sinusoidal case that may not reflect reality.

Example: L_∞ Norm

An eighth-order Chebyshev II filter

$$H(z) = 0.0888 \frac{\left(\begin{array}{c} z^8 + 4.43z^7 + 10.76z^6 + 17.46z^5 + 20.48z^4 + 17.46z^3 + \\ 10.76z^2 + 4.43z + 1 \end{array} \right)}{\left(\begin{array}{c} z^8 + 1.10z^7 + 1.97z^6 + 1.55z^5 + 1.22z^4 + 0.61z^3 + 0.24z^2 + \\ 0.061z + 0.008 \end{array} \right)}$$

is implemented using cascade, direct II, ladder/lattice, and normal architectures. The decision on whether to maintain the input scale at $k = 0.0883$ or $k = 1$ is a design choice. Maintaining $k = 0.08883$ will result in a unity passband gain. If $k = 1$, the passband gain increases to 11.257 (i.e., 1/0.08883). Once k is chosen, the DFT of the impulse response is measured at the output of each shift register (state location), and the maximum value is assigned to $H_i^{\max}$. A useful observation is to note that only one register needs to be tested for a direct II architecture since all registers are directly chained together. For cascade architectures, consisting of a collection of second-order direct II sections, only one of the two registers of each second-order section need be monitored. For $k = 0.08883$, the L_∞ data and coefficient register bounds are shown in Table 17.1 for cascade, direct II, ladder/lattice, and normal architectures. The integer word length I presumes that coefficients, states, input, and output share a common data format. The production of the state norms is shown in Figure 17.5 for cascade and direct II filter forms. The tabled data indicate that the cascade filter's integer bit field (i.e., I) actually is established by the filter coefficient dynamic range requirements. Setting aside the coefficient problem, it is noted that the next largest dynamic range requirements is established by the output term $\|y[k]\|_\infty = 1$.

TABLE 17.1.   Summary of the Derived L_∞ Data and Coefficient Bounds

Architecture	H_i^{max}	@ State	$|y[k]|_\infty$	Max. Coef.	Max. $< 2^I$	I
Cascade	0.461	[7, 8]	1	1.86	1.86	1
Direct II	0.992	[1, . . . , 8]	1	19.17	19.17	5
Lattice/ladder	0.992	[1, 2]	1	5.11	5.11	3
Normal (cascade)	0.297	[7, 8]	1	3.2	3.27	2

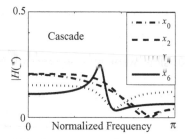 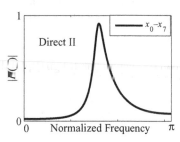

Figure 17.5 L_∞ state norms for cascade and direct II filters.

L_1 NORM BOUND

In the study of the L_2 state bound, it was noted that its practical use was compromised by the fact that many important inputs to a linear digital filter do not have a finite L_2 bound. The L_∞ frequency-domain state bound was found to be somewhat more practical, but suffered from the fact that it was based on a rigid steady-state sinusoidal assumption and that the input is a monotone sinusoidal signal. This too leaves the system vulnerable to potential run-time register overflows during transient periods, or in those instances where the input is not sinusoidal. What is desired is a state bound based on the individual samples of an arbitrary time series $v[k]$. It is reasonable to assume that the input sample values are bounded on a sample-by-sample basis. For example, the sample values leaving an analog-to-digital converter (ADC) may be assumed to be bounded by unity or $\|v[k]\|_\infty \le 1$. From Holder's inequality, if $p = \infty$, it follows that $q = 1$ since $1/p + 1/q = 1$. To guarantee the contents of the ith shift register bounded, the impulse response measured from input to the output of the ith shift register (i.e., $h_i[k]$) must be L_1 bounded. Unfortunately, computing the L_1 norm can be difficult in the general case. Specifically, the L_1 norm for the impulse response measured at the ith state location is

$$\|h_i\|_1 = \sum_{k=0}^{\infty} |h_i[k]|. \tag{17.13}$$

The L_1 norm of $h_i[k]$ was also called the worst case gain when applied to finite impulse responses (FIRs). If $h_i[k]$ is of infinite duration, computing the L_1 norm becomes more challenging. It is reasonable to assume, however, that the filter under study is stable.

In such cases, the impulse response $h_i[k]$ will eventually decay to some neighborhood of zero. The resulting trajectory can then be approximated by a finite time series. The advantage of estimating an L_1 bound can be appreciated when the worst case condition is considered. The worst case L_∞ input, having a unit bound, measured at the ith state location is $v[k] = \text{sign}(h_i[n - m])$. Convoluting the worst case input with the filter's impulse response, measured at state location "i," results in

$$y_i[k]_{\max} = \sum_{m=0}^{\infty} h_i[k - m]v[k] = \sum_{m=0}^{\infty} h_i[k - m]\text{sign}(h_i[k - m]) \le \|h_i\|_1 \|v\|_\infty = \|h_i\|_1, \quad (17.14)$$

which can be computed if $\|h_i\|_1$ can be approximated by a finite sum. That is,

$$\|h_i\|_1 = \sum_{k=0}^{\infty} |h_i[k]| \sim \sum_{k=0}^{N_1-1} |h_i[k]|, \quad (17.15)$$

where the value of N_1 is sufficiently large that the sum converges for all practical purposes.

Example: L_1 Bound Estimate

A sixth-order elliptic low-pass filter, with a -0.5-dB passband over $f \in [0, 5]$ kHz, -60-dB stopband over $f \in [7, 22.05]$ kHz, and a sample frequency $f_s = 44.1$ kHz, was designed. The filter's magnitude frequency response is displayed in Figure 17.6. The transfer function is

$$H(z) = \frac{\begin{array}{l}0.004 - 0.0034z^{-1} + 0.0079z^{-2} - 0.0036z^{-3} + 0.0079z^{-4} - \\ 0.0034z^{-5} + 0.004z^{-6}\end{array}}{\begin{array}{l}1 - 4.349z^{-1} + 8.8737z^{-2} - 10.0941z^{-3} + 6.8467z^{-4} - \\ 2.6188z^{-5} + 0.4416z^{-6}\end{array}}.$$

The sixth-order filter is factored into three second-order direct II filters having transfer functions of the form $H(z) = gH_1(z)H_2(z)H_3(z)$, $g = 0.004$, where

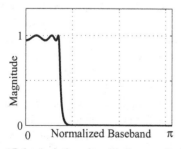
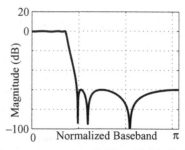

Figure 17.6   A sixth-order elliptic magnitude frequency response (left) and in decibels (right).

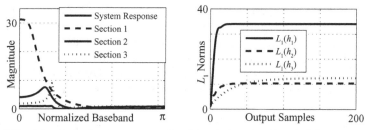

Figure 17.7 Magnitude frequency response of the sixth-order filter and three second-order sections (left). Also shown are the L_1 norms $\|h_i[k]\|_1$ of the output of each second-order section displayed over the first 200 output samples (right).

$$H_1(z) = \frac{1 + 0.9902z^{-1} + 1z^{-2}}{1 - 1.5132z^{-1} + 0.6109z^{-2}}$$

$$H_2(z) = \frac{1 - 0.7354z^{-1} + 1z^{-2}}{1 - 1.4645z^{-1} + 0.7746z^{-2}},$$

$$H_2(z) = \frac{1 - 1.1120z^{-1} + 1z^{-2}}{1 - 1.4552z^{-1} + 0.6331z^{-2}}.$$

For the sake of providing a common analysis forum, the input scale factor $g = 0.004$ is set to unity. The magnitude frequency response of the filter, measured at the output of each cascaded section is shown in Figure 17.7. The peak responses correspond to the L_∞ section norms in the frequency domain. The response of each second-order section can also be explored by monitoring the second-order filter's input–output impulse response $h_i[k]$, $i \in [1, 3]$, at state location $x_1[k]$ for each cascaded section. The stage-by-stage output impulse response can be computed and displayed in Figure 17.7. It can be seen that each trajectory converges to a steady-state value, but at differing rates. After about 200 samples the impulse responses have essentially converged. This information can be used to directly compute $\|h_i[k]\|_1$ for each second-order section. The L_1 norms $\|h_i[k]\|_1$ are determined to be $\|h_1[k]\|_1 = 34.04$, $\|h_2[k]\|_1 = 10.36$, and $\|h_3[k]\|_1 = 12.45$.

Suppose the input has a unit L_∞ norm, that is, $\|v[k]\|_\infty = 1$. Then the magnitude of the first section's output $|y_1[k]| \leq \|h_1[k]\|_1 \times 1 = 34.04$. The next output would be bounded by $|y_2[k]| \leq \|h_2[k]\|_1 \times |y_1[k]| = 10.36 \times 34.04 = 352.65$. The final stage output would be bounded by $|y_3[k]| \leq \|h_3[k]\|_1 \times |y_2[k]| = 12.45 \times 352.65 = 43905.5$ $4 < 2^{15.4}$. The analysis assumes that the input to all second-order filter sections are worst case inputs, an unrealistic assumption in many cases. At first glance, it would appear that 15+ bits of headroom would excessively stress a 16-bit processor. It should be remembered, however, that the L_1 method is very conservative and naturally overestimates the headroom requirement. Returning the scale factor $g = 0.004$ to the filter reduces $|y_3[k]| \leq 17.56 < 2^5$, a more realistic headroom requirement. Assuming that the L_1 norm is conservative, 3 or 4 bits of integer precision may be all that is needed to inhibit run-time register overflow. This analysis, however, does not account for internal dynamic range growth. This can be factored into the design by analyzing the computed internal state bounds in each of the second-order direct

TABLE 17.2. *L_1* **Norm Outcome Comparisons**

Section 1	Section 2	Section 3
$|y_1[k]| \leq 34.04$	$|y_2[k]| \leq 10.46$	$|y_3[k]| \leq 12.45$
$|x_{11}[k]| \leq 11.24$	$|x_{12}[k]| \leq 10.06$	$|x_{13}[k]| \leq 28.88$

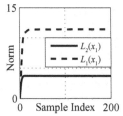 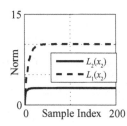 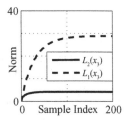

Figure 17.8 L_1 and L_2 state norms, shown for comparative purposes, showing the internal state register dynamics for three second-order direct II filter sections.

II sections. Since the two states of a second-order direct II filter are equally bounded, only one of the two impulse responses need to be computed for each section. The results are summarized in Table 17.2 and in Figure 17.8. The data are presented as the worst case outputs and state values per second-order section, assuming the input is unit bounded in an L_∞ sense (i.e., $\|v[k]\|_\infty = 1$). It can be seen that the internal headroom needed is similar to those predicted by analyzing worst case outputs.

NOISE POWER GAIN

It can be noted that a digital filter's dynamic range requirement and precision parameters are always in a state of tension. Fortunately, the trade-off between dynamic range and retained precision can often be mathematically managed. The difference between the ideal and computed outputs, namely, $e[k] = y_{\text{ideal}}[k] - y_{\text{finite}}[k]$, is attributed to finite word length effects. Imprecision of this type can be studied in terms of quantization errors. Quantization errors, such as those found at the output of an ADC, or due to coefficient rounding, or arithmetic rounding, are produced on a sample-by-sample basis. Individual quantization errors are normally reported to be a uniformly distributed random process having a zero mean and variance $\sigma^2 = Q^2/12$, where Q is the quantization step size. Arithmetic round-off errors can be assumed to be injected into the system at the point where rounding occurs as shown in Figure 17.9. In Figure 17.9, two multiply-accumulate (MAC) architectures are shown. The first is called multiply-round-accumulate (MRA) architecture that takes a full-precision product, rounds the product to a single-precision word, and then accumulates the outcomes. The second, called multiply-accumulate-round (MAR) architecture accumulates all the full-precision products (using extended precision registers), then rounds the final accumulation to a single-precision word. In some cases, the MAC architecture is established by the implementation technology; in other cases, it can be set by the filter designer under program control. It should be

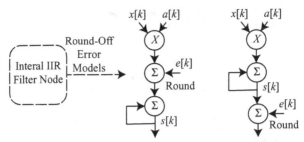

Multiply-Round-Add Multiply-Add-Round

Figure 17.9 Physical error model where $e[k]$ is a uniformly distributed random process representing round-off errors. The MRA architecture multiplies, rounds, then accumulates. The MAR architecture multiplies, accumulates, then rounds. The uncertainty in these processes is modeled by injecting quantization errors into the design at the points where rounding occurs.

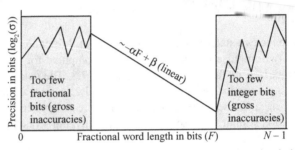

Figure 17.10 Typical IIR error variance versus fractional word length relationship showing high errors occurring when there are too many or too few fractional bits, and an ideal (linear) operating range.

appreciated that once an error is introduced into the system, it can perpetually stay within that system due to feedback. These recirculated errors can, in some instances, render the filter's output useless.

It should be recognized that a finite word length effect study of an IIR filter is substantially more difficult than an FIR filter. The fundamental difference is that noise introduced by arithmetic rounding can only be propagated along feed-forward paths for an FIR, whereas the noise injected into an IIR can be connected to a feedback path. As such, IIR rounding errors can recalculate and accumulate over time. The graph of the measured error variance versus the fractional word length for a typical IIR is suggested in Figure 17.10. Three operational regimes are delineated in this figure, and they correspond to the following possible modes of operation:

1. nonlinear mode (severe rounding) resulting in gross run-time errors introduced due to insufficient coefficient and data precision (i.e., F too small);

2. linear mode resulting in $\log(\sigma)$ precision versus the fractional word length F following the relationship that satisfies $\log(\sigma) \propto -\alpha F$;

3. nonlinear mode (saturation) resulting in gross run-time saturation overflows occur due to insufficient integer word length $I = N - F - 1$ (i.e., F too large).

Because of the complexity associated with the study of nonlinear systems, the error performance envelope of a system is often determined experimentally. Here, the arithmetic errors for a given IIR are computed over a collection of architectural and word length choices. A best design may then be selected in the context of minimizing the overall filter noise gain while simultaneously optimizing other performance parameters.

STATE-DETERMINED NOISE ANALYSIS

The study of errors within an IIR filter can be challenging. Consider first what happens to round-off noise injected into the ith shift register of an Nth-order IIR. Assume that the impulse response, measured between the input of the ith shift register and filter's output is computed and denoted $g_i[k]$. It should be fully appreciated that $g_i[k]$ includes the effects of all the feedback loops, which pass through the ith shift register. The noise injected into the ith shift register from all the attached rounded error sources will continuously recirculate within the filter while also being passed to the output. Therefore, the problem becomes one of measuring what happens to the noise over time. Statistically, the error variance measured at the output, due to a collection of internally injected noise sources attached to the ith shift register is given by

$$\sigma_i^2 = k_i \frac{Q^2}{12} \sum_{k=0}^{\infty} g_i[k] g_i^*[k] = k_i \frac{Q^2}{12} \|g_i\|_2^2, \tag{17.16}$$

where k_i reflects the number of independent round-off error sources, each having variance $Q^2/12$. The variance of the injected noise is scaled, or amplified by the path power gain $\|g_i\|_2^2$, existing between the ith register and output. Since there are N shift registers for an Nth-order canonic IIR architecture, the noise variance measured at the output is actually due to the collective contributions of all the amplified internal noise sources. That is, at the system-level, the error variance is

$$\sigma^2 = \sum_{i=1}^{N} \sigma_i^2 = \frac{Q^2}{12} \sum_{i=1}^{N} k_i \|g_i\|_2^2 = \frac{Q^2}{12} G^2. \tag{17.17}$$

The value σ^2 is the round-off error variance measured at the output and is seen to be the error variance of a single rounding source scaled by G^2, where G^2 is referred to as the filter's noise power gain. The rms value of G^2 (i.e., G) is called the noise gain. In addition, the form of Equation 17.17 is reminiscent of the computing formulas developed to support L_2 scaling policies as applied to state-determined IIR filters. The energy in this process, defined by Equation 17.17, can be represented in terms of the elements of an $N \times N$ matrix $W[k]$, a dyadic product given by

$$W[k] = g^{\mathrm{T}}[k] g[k] = (A^{\mathrm{T}})^{k-1} cc^{\mathrm{T}} (A)^{k-1}. \tag{17.18}$$

The on-diagonal elements of $W[k]$ correspond to $\|g_i[k]\|_2^2$. Fortunately, $\|g_i[k]\|_2^2$ can be computed using a general-purpose digital computer and using techniques developed earlier for L_2 analysis. More specifically, defining the matrix W to be

$$W = \sum_{k=0}^{\infty} W[k] = \sum_{k=0}^{\infty} g^T[k]g[k] = \sum_{k=0}^{\infty} \left(A^T\right)^k cc^T\left(A^T\right)^k. \tag{17.19}$$

It can be shown that W is a Lyaponov matrix satisfying $W = A^T W A + cc^T$. The Lyaponov matrix can be computed using the formula presented in Equation 17.9, which results in an error variance σ^2 and noise power gain G^2 formula given by

$$\sigma^2 = \frac{Q^2}{12} G^2 = \frac{Q^2}{12} \sum_{i=1}^{N} k_i W_{ii}. \tag{17.20}$$

It can be noted that everything required to compute Equation 17.20 is specified in terms of the elements of the state 4-tuple $[A, b, c, d]$, including k_i. The value of k_i can be directly equated to the number of noise sources in each feedback path. The determination of k_i can be automated by searching the ith row of the A matrix and counting the number of nonzero, nontrivial (± 1) coefficients in that row. Recall that if M multipliers are attached to the input of the ith shift register and rounded before being summed (MRA), then $k_i = M$ (see Fig. 17.9). If, however, all multipliers are maintained as full-precision products and are accumulated using an extended precision adder, and if the accumulator contents are then rounded, then $k_i = 1$ (see Fig. 17.9). Implicit in this model is the assumption that rounding the coefficients in $[c, d]$ will not appreciably add to the noise power gain.

Example: Noise Power Gain

An eighth-order Chebyshev II filter, having the transfer function

$$H(z) = 0.0888 \frac{\left(\begin{array}{c} z^8 + 4.43z^7 + 10.76z^6 + 17.46z^5 + 20.48z^4 + 17.46z^3 + \\ 10.76z^2 + 4.43z + 1 \end{array}\right)}{\left(\begin{array}{c} z^8 + 1.10z^7 + 1.97z^6 + 1.55z^5 + 1.22z^4 + 0.61z^3 + 0.24z^2 + \\ 0.061z + 0.008 \end{array}\right)}$$

is studied as direct II, cascade, and ladder/lattice filters. Each architecture has a unique state variable model and therefore a noise power gain. The noise power gains can be computed for a scaled and unscaled direct II, cascade, and lattice/ladder realization. Assuming an MRA architecture, the outcome is summarized in Table 17.3.

The noise power gains would indicate that all the architectures are essentially the same, operating at about 4.5 bits of lost precision. Over the suite of architectures reported, the cascade architecture is generally considered to be a good compromise. The actual statistical performance of a given architecture is a function of the complex interaction of arithmetic and coefficient round-off errors, which can, at various times,

TABLE 17.3. Eighth-Order IIR Example

Architecture	G^2	G	$\log_2(G)$
Cascade	594.56	24.37	<5 bits
Direct II	499.73	22.34	<5 bits
Lattice/ladder	705.86	26.57	<5 bits

(a)

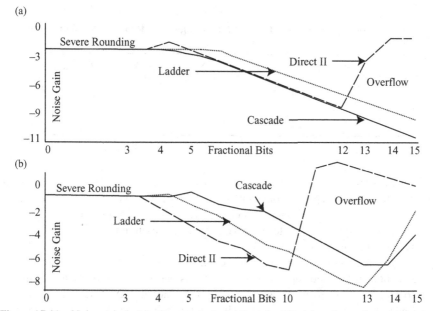

(b)

Figure 17.11 Noise gain in bits for a typical eighth-order IIR driven by an impulse (a) and a unit bound worst case input (b).

behave linearly or nonlinearly. Therefore, the noise models only provide a guideline of how an IIR may behave during run time.

The noise gain, in bits, can be experimentally studied by comparing the fixed-point system impulse response to the floating-point response, which is considered to be error free. The results of such an experiment are shown in Figure 17.11a for the cascade, direct II, and lattice/ladder cases, assuming the input is an impulse. The 16-bit architectures studied over all possible fractional word lengths $F \in [0, \ldots, 15]$. It can be seen that initially ($F \sim 0$), all the architectures suffer from severe coefficient quantization due to an insufficient fractional word width that causes small but critical coefficients, being rounded to zero. Here, for example, the input scale factor of $k = 0.08883$ will be mapped to zero for all $F \leq 3$ bits. For larger values of F, the system enters a linear mode of operation (no severe quantization or overflow errors). The measured round-off errors of the direct II and cascade are found to be nearly identical. The lattice/ladder response is similar, although of somewhat less precision. It should be remembered that the lattice/ladder was promoted as a low

TABLE 17.4. Noise Gain Performance of a Typical Eighth-Order IIR Using an Impulse Input

Type	Severe Rounding Range	Normal Operation	Overflow Saturation
Cascade	$F \in [0, \dots, 5]$	$F \in [6, \dots, 15]$	Not applicable
Direct II	$F \in [0, \dots, 4]$	$F \in [5, \dots, 12]$	$F \in [12, \dots, 15]$
Lattice/ladder	$F \in [0, \dots, 6]$	$F \in [7, \dots, 15]$	Not applicable

TABLE 17.5. Noise Gain Performance of a Typical Eighth-Order IIR with a Worst Case Input

Type	Severe Rounding Range	Normal Operation	Overflow Saturation
Cascade	$F \in [0, 6]$	$F \in [7, 13]$	$F \in [14, 15]$
Direct II	$F \in [0, 3]$	$F \in [4, 10]$	$F \in [11, 15]$
Lattice/ladder	$F \in [0, 5]$	$F \in [6, 13]$	$F \in [14, 15]$

coefficient round-off error sensitivity filter, which is not the same as claiming it to be immune to other types of round-off errors. At 12 bits, the direct II filter begins to suffer from a buildup of saturation errors, whereas the cascade and lattice/ladder filters performances continue to improve out to the maximum value of $F = 15$. The results are summarized in Table 17.4.

A problem associated with experimentally determining the effects of round-off noise is choosing a valid forcing function. This is particularly true when attempting to verify the scaled error variance predictions. Unexpected overflows can occur whenever the input is something other than a simple impulse. A unit step input $u[k]$ would be a more aggressive choice of test input for a low-pass filter. A popular choice is simply a uniformly distributed random noise $u[k] \in U[-1, 1]$ or linear chirp. The most aggressive and recommended input, however, is the worst case input. All these inputs have an L_∞ norm of unity but are internalized by the system differently. An example of this choice of forcing function is shown in Figure 17.11b and summarized in Table 17.5. It can be seen that in operation, the slope of each error trajectory is similar to that of the direct II, but the y-axis intercept values and duration of the linear descent differ. The linear operating range for the direct II is over the range $F \in [4, \dots, 10]$, $F \in [7, \dots, 13]$ for the cascade, and $F \in [6, \dots, 13]$ lattice/ladder. The lattice/ladder filter has a precision of about 8 fractional bits at $F = 13$. It is interesting to note that the optimal operating point of the lattice/ladder filter is $F = 13$ or $I = 2$, a point where the maximum lattice/ladder coefficient will be rounded from 5.1099 to 4. The coefficients can be seen to require at least $I = 3$ bits of integer precision. It can also be noted that the lattice/ladder superiority is present only over a subrange of F. For the simulation study, it would appear that the lattice/ladder, when set to $F = 13$, would provide the best statistical performance with the direct II ($F = 10$) and cascade ($F = 13$) behaving in manners similar to the lattice/ladder but about 2-bit inferior in precision. The maximum run-time bandwidth of the lattice/ladder would, however, normally be less than the direct II or cascade, which require fewer multiply-accumulations per filter cycle.

SIMILARITY TRANSFORMATION

It has been previously established that a system possessing a state variable model $[A, b, c, d]$ can be transformed into another, say $[\hat{A}, \hat{b}, \hat{c}, \hat{d}]$, using a linear transform T. It has been previously established that

$$\hat{A} = TAT^{-1}; A = T^{-1}\hat{A}T; \hat{b} = Tb; b = T^{-1}\hat{b}; \hat{c}^{\mathrm{T}} = c^{\mathrm{T}}T^{-1}; c^{\mathrm{T}} = \hat{c}^{\mathrm{T}}T; \hat{d} = d, \qquad (17.21)$$

which results in the following noise power gain matrix for the transformed system

$$W = A^{\mathrm{T}}WA + cc^{\mathrm{T}} = T^{\mathrm{T}}\hat{A}\left(T^{\mathrm{T}}\right)^{-1}WT^{-1}\hat{A}T + T^{\mathrm{T}}\hat{c}\hat{c}^{\mathrm{T}}T \qquad (17.22)$$

that can be simplified to read $\hat{W} = \left(T^{\mathrm{T}}\right)^{-1}WT^{-1}$ and is the noise power gain matrix for the new system. The significance of this result is that any change in architecture will normally produce a different noise power gain.

CHAPTER 18

INTRODUCTION TO MULTIRATE SYSTEMS

BACKGROUND

Digital filters accept sampled inputs and produce sampled outputs. In between, an input signal's time- and/or frequency-domain attributes are modified. Normally, the input sample rate f_{in} equals the output sample rate f_{out} (i.e., $f_s = f_{in} = f_{out}$). When they differ, a multirate solution arises. Multirate solutions are found in a wide range of applications, such as audio signal processing where various audio subsystems operate with different sample rates (e.g., 40 kHz vs. 44.1 kHz). In still other applications, multirate techniques can be used to reduce channel bandwidth and computational requirements. With the increased interest in multirate solutions, it is important that the design engineer be familiar with multirate design practice.

DECIMATION

If a time series $x[k]$ is accepted at a sample rate f_{in} and exported at a rate f_{out}, such that $f_{in} > f_{out}$, then the signal is said to be decimated* by M, where $M = f_{out}/f_{in}$. If M is an integer, then the decimated time series $x_d[k]$ is given by $x_d[k] = x[Mk]$, retaining only every Mth sample of the original time series, discarding all others. The effective sample rate is reduced from f_{in} to $f_d = f_{in}/M$ Sa/s. Formally, a decimated by M time series satisfies

$$x_d[n] = \sum_{k=-\infty}^{\infty} x[k]\delta(n - kM) \qquad (18.1)$$

with a z-transform given by $X_d(z) = X(z^M)$. The frequency signature of the decimated signal, relative to the undecimated parent, is given by

---

* Decimation originally referred to a disciplinary action employed by the Romans in dealing with mutinous soldiers. The mutineers were forced to select balls from an urn containing 10 times more white balls than black balls. The holders of black balls would be put to the sword. Therefore, every 10th soldier would be slain or decimated.

Digital Filters: Principles and Applications with MATLAB, First Edition. Fred J. Taylor.
© 2012 by the Institute of Electrical and Electronics Engineers, Inc.
Published 2012 by John Wiley & Sons, Inc.

$$X_d(e^{j\phi}) = X(e^{jM\phi}),\tag{18.2}$$

which is a frequency scaled version of the original signal spectrum. Furthermore, the decimated signal spectrum periodically repeats on f_s/M centers. This process is graphically shown in Figure 18.1.

One should be aware that Shannon's sampling theorem also applies to decimated signals. Suppose the highest frequency found in a time series $x[k]$ is B Hz. Aliasing can only be avoided if the decimated sample rate exceeds $f_d = 2B$ Hz. This means that there is a practical upper bound to the decimation rate. Referring to Figure 18.2, it can be seen that for unaliased decimation to take place, $f_s/M - B > M$ or $M > f_s/2B$. Increasing the decimation rate beyond this value will introduce the possibility of aliasing, as shown in Figure 18.2. In practice, the maximal decimation rate is rarely used. Instead, a more conservative oversampled rate is generally employed that will allow for a well-defined guard band as suggested in Figure 18.2.

The spectral behavior of a decimated band-limited signal can be studied graphically. Consider the band-limited signal shown in Figure 18.3. Decimation can be used to rearrange the spectrum and place copies of the signal spectrum at selected baseband locations. Suppose $(m)\pi/2M \leq \omega_s \leq (m + 1)\pi/2M$, where M is the decima-

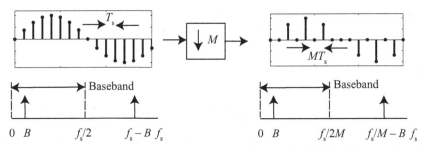

Figure 18.1 Magnitude spectrum of a parent and decimated by M ($M = 2$ shown) signals are interpreted. The input signal spectrum is assumed to be baseband limited to B Hz and the Nyquist frequency is $f_s/2$. The spectrum is periodically extended on multiples of the sample frequency. The decimated signal's spectrum is baseband limited to B Hz, as before, but the output Nyquist frequency is now $f_s/2M$ with sample frequency $f = f_s/M$. The spectrum is periodically extended on multiples of the new sample frequency.

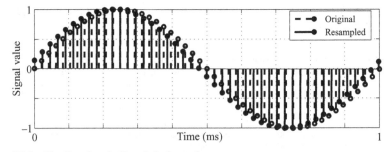

Figure 18.2 Unaliased and aliased decimated cases.

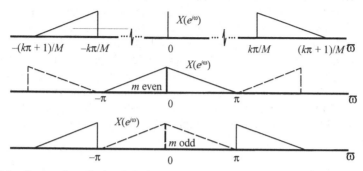

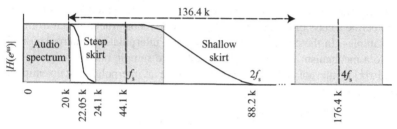

Figure 18.3   Decimation of a band-limited signal for *m* even and *m* odd showing different baseband symmetric outcomes.

Figure 18.4   Oversampled 4× commercial CD-ROM audio system.

tion rate and *m* is a positive integer. The resulting decimated spectrum is shown in Figure 18.3. If *m* is even, the original spectrum is translated down to 0 Hz (DC) ($\omega = 0$). If *m* is odd, the spectrum is a reflection of the original spectrum, also translated down to DC.

Example: Audio CD-ROM

It is normally assumed that a high-quality audio is band-limited to 20 kHz. This class of signal can be oversampled using an industry standard 44.1-kHz multimedia analog-to-digital converter (ADC). The nature of periodic sampling states that the baseband spectrum is replicated on integer multiples of 44.1 kHz. To insure that no aliasing occurs, an anti-aliasing analog filter is attached to the ADC's input. The purpose of the anti-aliasing filter is to remove signal energy found above the Nyquist frequency, such as those generated by the digital components, which are clocked at high rates. The rise and fall of these digital pulses generate high-frequency signals that can find their way to the ADC's input. To suppress such aliasing signals, the low-pass anti-aliasing analog filter would need a 20-kHz passband, and a stopband beginning no later than 24.1 kHz, resulting in a transition band of no more that $\Delta f = 4.1$ kHz, as shown in Figure 18.4. This is the specification for a very steep-skirt analog filter. Such narrow transition band analog filters are impractical when cost and packing requirements are considered. To achieve a viable solution, the design requirements need to be relaxed. To demonstrate the design differences between the unrelaxed and relaxed designs, refer to Figure 18.4, where the ±20-kHz baseband spectrum is displayed along with the shape of the anti-aliasing analog filter for $f_s = 44.1$ and 156.4 kHz. The choice of $f_s = 44.1$ kHz requires the use of a steep-skirt anti-aliasing analog filter, which is difficult to achieve in practice. For $f_s = 176.4$ kHz (4× oversampled), a much simpler low-order filter can serve as an analog anti-aliasing filter. This assumes that there is no appreciable signal energy from 20 kHz out to 156.4 kHz ([176.4 − 20] kHz). The analysis can be repeated for the case where $f_s = 48$ kSa/s, with a 4× oversampling rate of 192 kSa/s.

INTERPOLATION

The antithesis of decimation is called interpolation or upsampling. The use of the word interpolation is somewhat unfortunate since interpolation has previously been used to define a class of operations that are used to reconstruct a signal from a sparse set of samples. In the context of decimation and interpolation, interpolation simply refers to a mechanism that increases the effective sample rate of a signal. Suppose a signal $x[k]$ is interpolated by a factor N to create an interpolated or upsampled time series $x_i[k]$, where

$$x_i[k] = \begin{cases} x[k] \text{ if } k = 0 \bmod N \\ 0 \quad \text{otherwise} \end{cases}$$ (18.3)

The act of interpolation by a factor N is facilitated by inserting $N - 1$ zeros in between the adjacent samples of the original time series. This action is sometimes referred to as zero padding. The result is a time series sampled at a rate f_{in} that is interpolated into a new time series sampled at the elevated sample rate $f_{out} = Nf_{in}$.

Interpolation is often directly linked to decimation. To illustrate, suppose $x_d[k]$ is decimated by the M version of a time series $x[k]$, which was sampled at a rate f_s. Then $x_d[k]$ contains only every Mth sample of $x[k]$ and is defined with respect to a decimated sample rate $f_d = f_s/M$. Interpolating $x_d[k]$ by N would result in a time series $x_i[k]$, where $x_i[Nk] = x_d[k]$ and 0 otherwise. The sample rate of the interpolated signal would therefore be increased from f_d to $f_i = Nf_d = Nf_s/M$. If $N = M$, the output sample rate would be restored to f_s. It can be noted that $x_i[k] \neq x[k]$ due to the loss of sample values during the decimation process. If $x[k]$ is first interpolated by N and then decimated by N, $x[k] = x_d[k]$.

Relative to the decimated signal $x_d[k]$, the frequency-domain signature of an interpolated by N signal $x_i[k]$ can be defined in terms of the z-transform of Equation 18.3, which states that

$$X_i(e^{j\omega}) = X(e^{jN\omega}).$$ (18.4)

Therefore, the frequency-domain representation of the zero-padded or interpolated signal is that of the original signal's spectrum replicated in the frequency domain as shown in Figure 18.5. As predicted by Equation 18.4, the presence of the complex

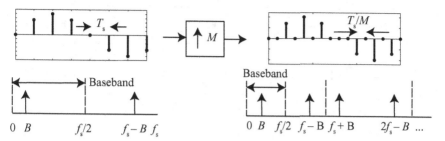

Figure 18.5   Frequency response of a zero-padded interpolated signal (shown for $M = 2$).

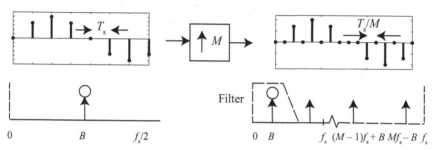

Figure 18.6   Spectra of decimated signal, zero-filled interpolated signal, and filtered baseband signal. Also shown is an overlay of an interpolation filter.

exponential $e^{jN\omega}$ gives rise to periodic replications of the spectrum on centers that are integer multiples of the original sample rate.

For a given integer N, the frequency response of the interpolated signal can be computed and compared with that of the original time series. The result of interpolating by N a low-frequency signal sampled at a rate f_s/N is displayed in Figure 18.6. Observe that the resulting interpolated spectrum contains copies of the baseband spectrum located on f_s/N frequency centers. The signal spectra found at the output of an interpolator is seen to contain multiple copies of the baseband spectrum. The unwanted copies generally need to be removed before the interpolated signal can be made useful. If the interpolated signal is first converted into an analog domain using a digital-to-analog converter (DAC), a simple RC low-pass circuit can sometimes be used to eliminate the unwanted extraneous spectral components (i.e., images). If the elimination of the unwanted copies of the baseband spectrum from the interpolated spectrum is to be performed digitally, an ideal Shannon interpolating filter is known to be optimal, but also impractical. As a result, the ideal Shannon interpolation filter is generally approximated by a practical low-pass filter having a passband covering the frequency range $f \in (-f_s/2N, f_s/2N)$.

SAMPLE RATE CONVERSION

A commonly encountered signal processing problem is interfacing two systems having dissimilar sample rates, say f_{in} and f_{out}. This defines a sample rate conversion

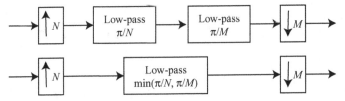

Figure 18.7 Basic rate conversion system.

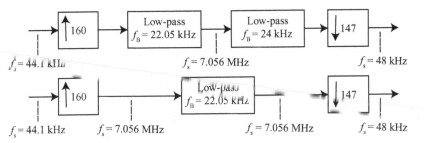

Figure 18.8 Sample rate conversion example.

problem. If the ratio of f_{in} to f_{out}, or vice versa, is a rational fraction, then direct decimation and interpolation can be used. To illustrate, suppose $k = N/M$ where N and M are integers where $f_{in} = kf_{out}$. The non-integer sample rate converter system is described in Figure 18.7. The indicated filters, whether separate or combined, either "clean up" an interpolated signal or provide anti-aliasing services for the decimator.

Example: Sample Rate Conversion System

Two typical audio sampling rates are 44.1 kSa/s (multimedia) and 48 kSa/s (audio tape). It follows that the conversion ratio $k = 48,000/44,100 = 160/147$ can be reduced no further. The implication is that CD sample rate of 44.1 kSa/s will need to be interpolated by a factor of 160 to a sample rate of 7.056 MSa/s. This requires that the upsampler, or interpolator's output be elevated to a 7.056 MSa/s rate as shown in Figure 18.8. The low-pass interpolating filter has a cut-off frequency around 22.05 kHz to insure that energy above the input Nyquist frequency is eliminated. The anti-aliasing low-pass filter, located before the decimator must have a bandwidth no greater than 24 kHz based on an ultimate 48-kHz output. The two low-pass interpolating and aliasing filters, when cascaded, would result in a 22.05-kHz filter (see Fig. 18.8). There is, however, a potential problem with implementing sample rate conversion at high interpolation and decimation rates. Connecting a 44.1 kSa/s device to a 40 kSa/s unit would, for example, require even higher interpolating and decimation rates, specifically 400 and 441. This may be impractical in some cases. In such cases, it is sometimes easier to send the 44.1 kSa/s signal to a DAC, convert it into an analog signal, and then resample the analog signal at 48 kSa/s.

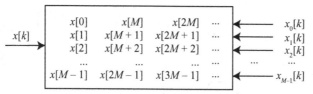

Figure 18.9   Block decomposition of the signal space.

POLYPHASE REPRESENTATION

Interpolated and decimated signals and systems can be studied in a piecemeal fashion. However, a more robust and formal means of mathematical modeling multirate systems is called a polyphase decomposition. Polyphase models are defined in the z-domain and are used to represent an arbitrary time series $x[k]$, sampled at a rate of f_s Sa/s. Initially assume that a discrete-time signal $x[k]$ has a z-transform representation given by $X(z) = \sum x[k]z^{-k}$. Suppose the time series $x[k]$ is partitioned into M distinct data sequences as shown in Figure 18.9 (called block decomposition).

The ith block can be seen to be the original time series delayed by "i" samples and decimated at a rate M. In terms of a z-transform, the block decomposed time series can be expressed as

$$X(z) = \begin{cases} \dots (x[0] + z^{-M}x[M] + z^{-2M}x[2M] + \dots) \\ \dots z^{-1}(x[1] + z^{-M}x[M+1] + z^{-2M}x[2M+1] + \dots) \\ \dots \qquad \dots \qquad \dots \qquad \dots \qquad \dots \\ \dots z^{-(M-1)}(x[M-1] + z^{-M}x[2M-1] + z^{-2M}x[3M-1] + \dots) \end{cases}. \qquad (18.5)$$

The ith row of Equation 18.5 can now be defined in terms of the ith polyphase term $P_i(z)$ that is given by

$$X(z) = \sum_{i=0}^{M-1} z^{-i} P_i(z^M); \; P_i(z) = \sum_{k=-\infty}^{\infty} x(kM+i)z^{-k}. \qquad (18.6)$$

The resulting compact representation is called an M-component polyphase decomposition of the time series $x[k]$. A multirate system can also be described in terms of the transposed polyphase function denoted $Q_i(z)$, which is related to the polyphase function $P_i(z)$ through

$$X(z) = \sum_{i=0}^{M-1} z^{-(M-1-i)} Q_i(z^M); \; Q_i(z) = P_{M-1-i}(z). \qquad (18.7)$$

The mechanics of a polyphase and transpose polyphase decomposition are summarized in Figure 18.10. Notice that for the case where $M = 4$, the solution consists of four parallel channels along with $M - 1 = 3$ shift registers to properly phase the signals. Each channel consists of a 4:1 decimator and a path from input to output, which is clocked at $f_s/4$. The bandwidth requirements of each individual channel is $1/M = 1/4$th that of a direct input/output path. The interleaved polyphase components can then be recombined to reconstruct the original signal at the original sample rate.

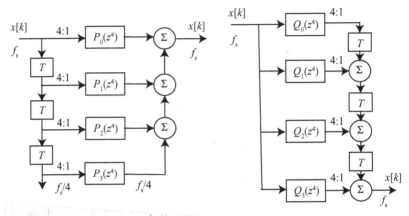

Figure 18.10   Example $M = 4$ polyphase multirate system

A polyphase representation can also be used to examine the physical act of decimation. Consider a time series $x[k]$, which is decimated by a factor M to produce a new time series $x_d[k]$, which is given by $x_d[k] = x[Mk]$. Suppose further that the z-transform of $x[k]$ is known for an arbitrary time series $x[k]$ and is given by $X(z) = \Sigma\ x[k]z^{-k}$. With respect to a sample frequency f_s, each sample delay (i.e., z^{-1}) has a duration of $T_s = 1/f_s$. The z-transform of the decimated time series can then be expressed as $X_d(z) = \Sigma\ x_d[k]z^{-k} = P_0(z)$. It is important to notice that each delay now has a duration based on a decimated clock rate having a period $T_s = M/f_s$. After some algebraic manipulation, $X_d(z)$ can be reinterpreted as

$$X_d(z) = P_0(z) = \frac{1}{M}\sum_{k=0}^{M-1} X(W_M^k z^{1/M}); \; W_M = e^{-j2\pi/M}, \tag{18.8}$$

which can be verified by performing the summation term-by-term to achieve

$$\frac{1}{M}\sum_{k=0}^{M-1} X(W_M^k z^{1/M})$$

$$= \frac{1}{M}\left(X[W_M^0 z^{1/M}] + X\left[W_M^1 z^{1/M}\right] + X\left[W_M^2 z^{1/M}\right] + \ldots + X\left[W_M^{M-1} z^{1/M}\right]\right)$$

$$= \frac{1}{M}\left\{ \begin{array}{c} x[0] + x[1]z^{-1/M} + x[2]z^{-2/M} + \ldots + x[M]z^{-M/M} + \ldots \\ + x[0] + x[1]W_M^1 z^{-1/M} + x[2]W_M^2 z^{-2/M} + \ldots + x[M]W_M^M z^{-M/M} + \ldots \\ \ldots \quad\quad\quad \ldots \\ + x[0] + x[1]W_M^{M-1} z^{-1/M} + x[2]W_M^{2(M-1)} z^{-2/M} + \ldots + x[M]W_M^{M(M-1)} z^{-M/M} + \ldots \end{array} \right\}$$

$$= \frac{1}{M}\left\{ Mx[0] + 0x[1]z^{-1/M} + 0x[1]z^{-2/M} + \ldots + Mx[M]z^{-M/M} + \ldots\right\}$$

$$= \frac{1}{M}\left(Mx[0] + Mx[M]z^{-M/M} + \ldots\right) = P_0(z). \tag{18.9}$$

The term z^{-1} refers to a single clock delay relative to the decimated sample rate $f_d = f_s/M$, or equivalently a delay of $T_d = M/f_s$ with respect to the input sample rate. Therefore, $z^{-1/M}$ physically corresponds to a delay of $T_d/M = T_s = 1/f_s$.

Example: Polyphase Spectrum

The time series $x[k] = a^k u[k]$, where $|a| < 1$, represents a decaying exponential signal being sampled at a rate of f_s samples per second. Assume that two decimation rates, namely, $M = 2$ and $M = 4$, are to be tested. From a standard table of z-transforms, it follows that the z-transform of the original time series, clocked at f_s, is $X(z) = 1/(1 - az^{-1})$ where z^{-1} corresponds to a delay of $T_d = 1/f_s$ seconds. The decimated by 2 signal has a time series given by $x_d[k] = x[2k] = \{1, a^2, a^4, \ldots\}$. The z-transform of this decimated signal satisfies $X(z) = 1/(1 - a^2 z^{-1})$, where z^{-1} now corresponds to the delay of the decimated signal, which is given by $T_d = 2/f_s$ seconds. From Equation 18.8 and $M = 2$, it follows that for $W_2^0 = 1$, $W_2^1 = -1$:

$$X_d(z) = \frac{1}{2}\left(X\left(W_2^0 z^{1/2}\right) + X\left(W_2^1 z^{1/2}\right)\right),$$

where, at the decimated sample rate of $f_d = f_s/2$, the reconstructed signal components are

$$X(W_2^0 z^{1/2}) = X(z^{1/2}) = \sum_{k=0}^{\infty} a^k z^{-k/2} = \frac{1}{1 - az^{-1/2}} \leftrightarrow \{1, a, a^2, a^3, \ldots\},$$

$$X(W_2^1 z^{1/2}) = X(-z^{1/2}) = \sum_{k=0}^{\infty} (-1)^{-k/2} a^k z^{-k/2} = \frac{1}{1 + az^{-1/2}} \leftrightarrow \{1, -a, a^2, -a^3, \ldots\}.$$

It can immediately be seen that upon combining these terms, $X_d(z)$ and $x_d[k]$ result. For $M = 4$, $x_d[k] = x[4k] = \{1, a^4, a^8, \ldots\}$ and $X_d(z) = (1/4)\left(X\left(W_4^0 z^{1/4}\right) + X\left(W_4^1 z^{1/4}\right) + X\left(W_4^2 z^{1/4}\right) + X\left(W_4^3 z^{1/4}\right)\right)$, where, at the sample rate of $f_d = f_s/4$, the individual delay terms are defined with respect to f_s and satisfy

$$X\left(W_4^0 z^{1/4}\right) = X\left(z^{1/4}\right) = \sum_{k=0}^{\infty} a^k z^{-k/4} = \frac{1}{1 - az^{-1/4}} \leftrightarrow \{1, a, a^2, a^3, \ldots\},$$

$$X\left(W_4^1 z^{1/4}\right) = X\left(-jz^{1/4}\right) = \sum_{k=0}^{\infty} (-j)^{k/4} a^k z^{-k/4} = \frac{1}{1 + jaz^{-1/4}} \leftrightarrow \{1, -ja, -a^2, ja^3, \ldots\},$$

$$X\left(W_4^2 z^{1/4}\right) = X\left(-z^{1/4}\right) = \sum_{k=0}^{\infty} (-a)^k z^{k/4} = \frac{1}{1 + az^{-1/4}} \leftrightarrow \{1, -a, a^2, -a^3, \ldots\},$$

$$X\left(W_4^3 z^{1/4}\right) = X\left(jz^{1/4}\right) = \sum_{k=0}^{\infty} (j)^{k/4} a^k z^{-k/4} = \frac{1}{1 - jaz^{-1/2}} \leftrightarrow \{1, ja, -a^2, -ja^3, \ldots\}.$$

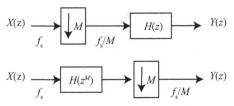

Figure 18.11   Equivalent decimated systems.

Again it can immediately be seen that upon combining these terms, $X_d(z)$ and $x_d[k]$ result.

A logical question to pose relates to the location of the decimator in a typical signal processing stream. The two systems shown in Figure 18.11 are functionally equivalent. The equivalence of the two architectures is sometimes referred to as the noble identity. The topmost path consists of a decimator and a filter. The bottom path consists of a filter, which is identical to that found in the top loop, except its clock is running M times faster than the rate that is found in the top loop. Both designs have the same number of coefficients and therefore the same arithmetic complexity. The major difference between the circuits is found in the rate at which the coefficient multiplications must be performed. The topmost filter has an internal data rate 1/Mth that of the bottom filter. Therefore, the top architecture is generally preferred due to its lower real-time computational requirement.

Example: Polyphase Filter Description

Consider passing a discrete-time signal through a finite impulse response (FIR) filter given by $H(z) = 2 + 3z^{-1} + 3z^{-2} + 2z^{-3}$, followed by the decimation of the filter's output by $M = 2$. The filter is implemented using the polyphase description shown in Figure 18.12. For $M = 2$, Equation 18.6 states that $H(z) = P_0(z^2) + z^{-1}P_1(z^2) = [2 + 3z^{-2}] + z^{-1}[3 + 2z^{-2}]$ or $P_0(z^2) = 2 + 3z^{-1}$; $P_1(z^2) = 3 + 2z^{-1}$. The fourth-order FIR is now represented as two interleaved second-order FIRs. While the circuits shown in Figure 18.12 are equivalent, the data being filtered by Circuit A arrive at the polyphase filter's input at a rate half of that seen by Circuit B. Therefore, Circuit A would have the lowest arithmetic bandwidth demands (multiply-accumulate [MAC]/s) of the two choices.

A detailed analysis of the filtering process would indicate that at various sample instances, the filter's intermediate values are

$$x_a[k] = \{\ldots, x[0], x[2], x[4], x[6], \ldots\},$$
$$y_a[k] = \{\ldots, 2x[0]+3x[-2], 2x[2], +3x[0], 2x[4]+3x[2], 2x[6]+3x[4], \ldots\},$$
$$x_b[k] = \{\ldots, x[1], x[3], x[5], x[7], \ldots\},$$
$$y_b[k] = \{\ldots, 3x[1]+2x[-1], 3x[3], +2x[1], 3x[5]+2x[3], 3x[7]+2x[5], \ldots\}.$$

At time sample $k = 4$, $y[4] = 2x[4] + 3x[3] + 3x[2] + 2x[1] = y_a[3] + y_b[2]$.

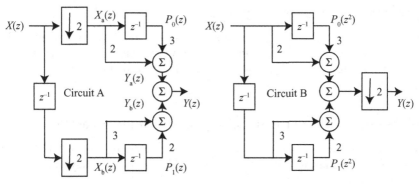

Figure 18.12 Equivalent multirate filters.

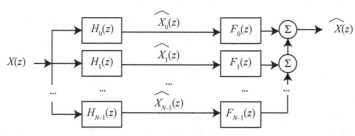

Figure 18.13 Typical sub-band decomposition system showing the analysis (left), and synthesis (right) filters. The output sample rate of the analysis filters is $1/M$th of the input sample rate. The output sample rate of the synthesis filter is M times higher than their input rate.

SUB-BAND FILTERS

Multirate systems often appear as a bank of filters, where each filter maps the input into a sub-band. Refer to Figure 18.13 that separates the filters into a multirate subsystem classified as analysis filter and synthesis filter systems. The analysis filters, denoted $H_i(z)$, convolves the input signal into M sub-bands. The synthesis filters, denoted $F_i(z)$, reconstruct the signal from the sub-band components. This structure is common to many bandwidth compression and signal and image enhancement applications. Sub-band filters are designed so that the data rate requirements of each channel, shown in Figure 18.13, are rated to be $1/M$th that of the input. This allows low data rate channels to be used to communicate analysis data to the synthesizer for reconstruction.

MATLAB

MATLAB contains a number of multirate infrastructure support tools. They are discussed in the following sections.

Decimation: Decrease Sample Rate

Operation $y = \texttt{decimate(x,r)}$ reduces the sample rate of x by a factor r. The decimate operation automatically includes a low-pass anti-aliasing filter having a passband ranging from $f \in [0, 0.4f_s]$. Figure 18.14 displays a decimation by 4 operation.

Downsample: Decrease Sampling Rate by Integer Factor

Operation $y = \texttt{downsample(x,n)}$ decreases the sampling rate of x by keeping every nth sample starting with the first sample. Figure 18.15 displays a decimation by 4 operation.

Interp: Increase Sampling Rate by Integer Factor

Operation $y = \texttt{interp(x,r)}$ increases the sampling rate of x by a factor of r. Figure 18.16 displays an interpolation by 4 operation.

Interp1: 1-D Data Interpolation (Table Lookup)

Operation $\texttt{yi} = \texttt{interp1(x,Y,xi)}$ interpolates to find yi, the values of the underlying function Y at the points in the vector or array xi. A number of interpolation schemes are supported. Here, x must be a vector. Figure 18.17 displays the interpolation of sine wave.

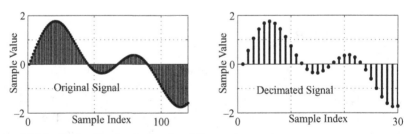

Figure 18.14   Decimation by 4 of an input time series.

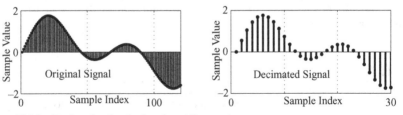

Figure 18.15   Decimation by 4 of an input time series.

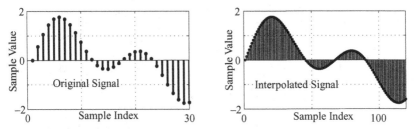

Figure 18.16   Interpolation by 4 of an input time series.

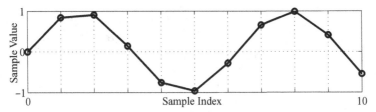

Figure 18.17   Interpolation of a sine wave using Interp1.

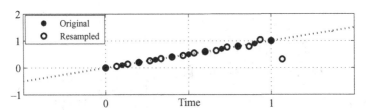

Figure 18.18   Sample rate conversion of 3/2.

Resample: Change Sampling Rate by Any Rational Factor

Operation $y = \texttt{resample(x,p,q)}$ resamples the sequence in vector x at p/q times the original sampling rate, using a polyphase filter implementation. Figure 18.18 displays a sample rate conversion by 3/2.

Upsample: Increase Sampling Rate by Integer Factor

Operation $y = \texttt{upsample(x,n)}$ increases the sampling rate of x by inserting $n - 1$ zeros between samples. Figure 18.19 displays an upsampling by a factor 3.

Spline: Cubic Spline Data Interpolation

Operation $yy = \texttt{spline(x,Y,xx)}$ uses a cubic spline interpolation to find yy, the values of the underlying function Y at the values of the interpolate xx. For the interpolation, the independent variable is assumed to be the final dimension of Y with the breakpoints defined by x. Figure 18.20 displays the interpolation of a sine wave.

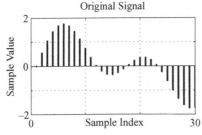 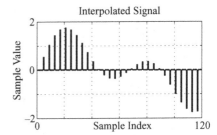

Figure 18.19   Upsampling by a factor 4.

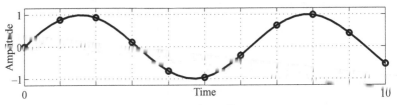

Figure 18.20   Interpolation of a sine wave using spline.

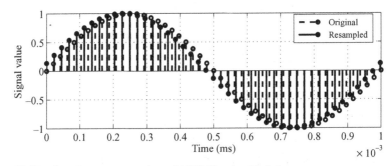

Figure 18.21   Sample rate conversion of 147/160 using Upfirdn.

Upfirdn: Upsample, Apply FIR Filter, and Downsample

Operation `yout = upfirdn(xin,h,p,q)` filters the input signal *xin* using FIR filter having an impulse response *h*, for *p* being the upsample value, *q* the downsample value (default $p = q = 1$). Figure 18.21 displays a change in the sampling rate by a factor of 147/160 (48 kHz [DAT rate] to 44.1 kHz [CD sampling rate]) and filters the data with an FIR filter.

MULTIRATE FILTERS

INTRODUCTION

Fixed clock rate filters represent today's mainstream filter technology. There are times, however, when such filters become excessively complex, especially when attempting to satisfy demanding performance specifications. In such cases, multirate filters can sometimes mitigate this problem, reducing complexity to an acceptable level. Over time, a number of multirate filter structures have emerged that are known to provide viable solutions in specific application instances.

DISCRETE FOURIER TRANSFORM (DFT) FILTER BANK

Broadband signals can often be decomposed into frequency restrictive sub-bands. An interesting application of this principle is called the uniform DFT filter bank, or DFT filter bank. A DFT filter bank has a magnitude frequency response suggested in Figure 19.1 and consists of a collection of identically shaped filters whose center frequencies are uniformly distributed across the baseband. The nth filter of an Mth order DFT filter bank, denoted $H_n(z)$, is defined in terms of the profile of a low-pass filter $H_0(z)$, called the prototype filter. A DFT filter bank translates the prototype's low-pass frequency response to new center frequencies $f_n = nf_s/M$, producing $H_n(z) = H_0(W_M^n z)$ for $n \in [0, M-1]$. The complex exponential term W_M^n performs a modulation service, mixing the prototype's impulse response, $h_0[k]$, with W_M^{nk} in order to perform the frequency translation. In the frequency domain, the nth subfilter's frequency response is therefore given by $H_n(e^{j\omega}) = H_0(e^{j(\omega - 2n\pi/M)})$. The result is a bank of frequency translated frequency-selective filters as shown in Figure 19.1 for $M = 8$.

Consider the case where the prototype filter $H_0(z)$ has a polyphase representation

$$H_0(z) = \sum_{i=0}^{M-1} z^{-i} P_{0i}(z^M),$$

(19.1)

Digital Filters: Principles and Applications with MATLAB, First Edition. Fred J. Taylor.
© 2012 by the Institute of Electrical and Electronics Engineers, Inc.
Published 2012 by John Wiley & Sons, Inc.

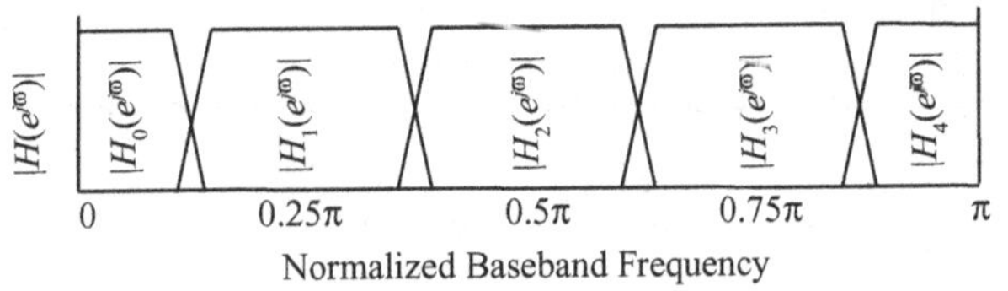

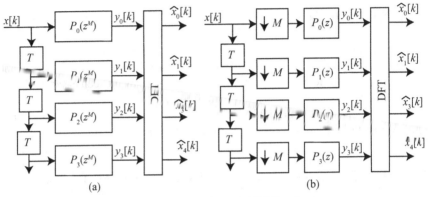

Figure 19.1 Uniform DFT filter bank magnitude frequency response for $M = 8$.

Figure 19.2 (a) DFT filter bank and (b) decimated DFT filter bank.

then

$$H_n(z) = \sum_{i=0}^{M-1} W_M^{-in} z^{-i} P_{0i}(W_M^{Mn} z^M) = \sum_{i=0}^{M-1} W_M^{-in} z^{-i} P_{0i}(z^M), \qquad (19.2)$$

which has the structure of a DFT (i.e., $\text{DFT}(x[k]) = X[n] = \sum W^{nk} x[k]$). When the polyphase outputs are combined using an M-sample DFT, as shown in Figure 19.2a, an M-band filter results. The complexity of a DFT filter bank can be analyzed in the context of the complexity of a filter and DFT. Adding the decimate by M circuit, shown in Figure 19.2b, reduces the multiply rate count by a factor of $1/M$. The multiplicative complexity of an M-point DFT can, in practice, be made small on the order of $M\log_2(M)$ if a radix-2 fast Fourier transform (FFT) can be used. MATLAB, it can be noted, does not contain direct DFT filter bank design support.

Example: DFT Filter Bank

A DFT filter bank is defined in terms of a low-pass prototype filter. Suppose the prototype finite impulse response (FIR) has a transfer function $H_0(z) = 2 + 3z^{-1} + 3z^{-2} + 2z^{-3}$. For $M = 2$, the polyphase filter representation is $H_0(z) = P_{00}(z^2) + z^{-1}P_{01}(z^2)$, where $P_{00}(z) = 2 + 3z^{-1}$, and $P_{01}(z) = 3 + 2z^{-1}$. Noting that $W_2^0 = 1$ and $W_2^{-1} = -1$ for a 2-point DFT, it follows that

$$H_0(z) = \sum_{i=0}^{1} z^{-i} P_{0i}(z^2) = P_{00}(z^2) + z^{-1} P_{01}(z^2) = 2 + 3z^{-1} + 3z^{-2} + 2z^{-3},$$

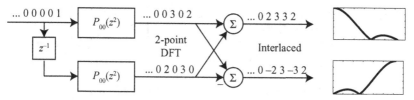

Figure 19.3   DFT filter bank and impulse response for $M = 2$.

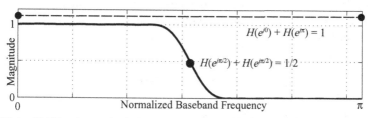

Figure 19.4   Half-band magnitude frequency response exhibiting midbase band symmetry.

$$H_1(z) = \sum_{i=0}^{1} W_2^{-i} z^{-i} P_{0i}(z^2) = P_{00}(z^2) - z^{-1} P_{01}(z^2) = 2 - 3z^{-1} + 3z^{-2} - 2z^{-3}.$$

The impulse response of the DFT filter bank is interpreted in Figure 19.3. The impulse response, measured along the top path, is seen to be equal to $\{2, 3, 3, 2\}$, which corresponds to the low-pass filter response $h_0[k]$. The impulse response measured along the bottom path is equal to $\{2, -3, 3, -2\}$, which corresponds to the high-pass filter response $h_1[k]$.

L BAND FILTERS (REVISITED)

An L band filter, also called a Nyquist filter, is odd order and characterized by $h[Lk] = 1/L$ for $k = 0$, 0 otherwise. That is, the 0th coefficient is $1/L$, but all others that are a multiple of L have a value of zero. This has implications in a polyphase representation of a filter. A half-band filter has about half of its coefficients equal to zero. The MATLAB function firhalfband found in the Filter Design Toolbox can be used to create an equiripple half-band FIR. Otherwise, half-band filters can be created using an equiripple FIR design tool. A half-band FIR has a baseband point of even symmetry at $f = f_s/4$.

Example: Half-Band Filter

A half-band filter ($L = 2$) has an impulse response $h[k]$ having a center-tap coefficient of ½ and all other even-indexed coefficients are zero. This means that about half of the filter's coefficients are zero. The frequency response of a typical half-band filter is shown in Figure 19.4. A half-band filter has known symmetry about the

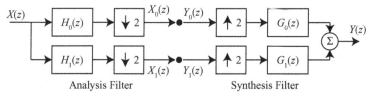

Figure 19.5   QMF filter architecture.

normalized frequency $f = f_s/4$ and is based on the L band filter theory. It also follows that $H(e^{j\phi}) + H(e^{j(\phi-\pi)}) = 1$ for any ϕ. Half-band filters will be used to develop the next class of filter.

QUADRATURE MIRROR FILTER (QMF)

Multirate systems are often used to reduce the required sample rate to a value that can be accepted by a band-limited communication channel. QMFs are a popular means of performing a sub-band signal decomposition so that the individual sub-bands can be processed through channels of reduced bandwidth. The basic architecture of a two-channel QMF system is shown in Figure 19.5. The two-channel QMF system establishes two input–output paths, each having a bandwidth requirement that is essentially half the original bandwidth requirements. Using this technique, a 2^M-channel QMF can be designed using an M-level binary tree architecture. The top path of the $M = 2$ solution shown in Figure 19.5 contains a low-pass filter, and the bottom path a high-pass filter. The spectral evolution of the signal elements of the system shown in Figure 19.5 is presented in Figure 19.6. Observe that the signals $x_0[k]$ (viz: $X_0[z]$) and $x_1[k]$ (viz: $X_1[z]$) are decimated by 2 after being filtered. The signals entering the decimators are $H_0(z)X(z)$ and $H_1(z)X(z)$, respectively. From the theory of polyphase representation, it follows that

$$X_0(z) = \frac{1}{2}\left\{X(z^{1/2})H_0(z^{1/2}) + X(-z^{1/2})H_0(-z^{1/2})\right\},$$

$$X_1(z) = \frac{1}{2}\left\{X(z^{1/2})H_1(z^{1/2}) + X(-z^{1/2})H_1(-z^{1/2})\right\}. \tag{19.3}$$

The signals $x_0[k]$ and $x_1[k]$ are then transmitted by the analysis filter section at the decimated rate ($f_s/2$), and carried to the synthesis filter section within two distinct channels. The signals are recovered by the synthesis filter section, which, upon postprocessing, restores the original signal at the original sample rate f_s. Interpreting $Y_0(z^2)$ and $Y_1(z^2)$, as shown in Figure 19.6, $Y(z)$ can be expressed as

$$Y(z) = G_0(z)Y_0(z^2) + G_1(z)Y_1(z^2)$$

$$= \frac{1}{2}\left\{(H_0(z)G_0(z) + H_1(z)G_1(z))X(z)\right\} + \frac{1}{2}\left\{(H_0(-z)G_0(z) + H_1(-z)G_1(z))X(-z)\right\}.$$

$\quad\quad \{\text{alias} - \text{free terms}\} \quad\quad\quad\quad\quad \{\text{aliased terms}\} \quad\quad\quad\quad (19.4)$

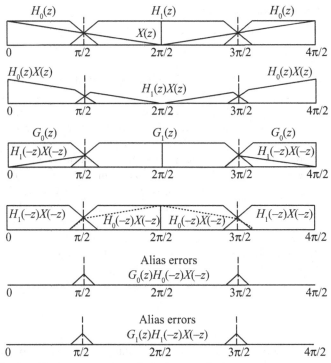

Figure 19.6   Graphical interpretation of aliasing in a QMF.

The problem, however, is that in the frequency domain, the responses found in the two transmission channels partially overlap. If left uncorrected, the overlapping signals could introduce aliasing errors into the final solution. A possible aliasing suppression strategy is graphically shown in Figure 19.6. The aliasing error potential can be suppressed if ideal (boxcar) filters are assumed, but that is impractical. Consider instead the use of only physically realizable filters to remove potential aliasing effects. The first term, or factor, found in Equation 19.4 contains all the alias-free information needed to reconstruct the input signal. The second term contains all the alias terms that need to be removed for alias-free signal restoration. To suppress aliasing effects, it is required that

$$\{(H_0(-z)G_0(z) + H_1(-z)G_1(z))\, X(-z)\} = 0, \tag{19.5}$$

which is trivially satisfied if $G_0(z) = H_1(-z)$ and $G_1(z) = -H_0(-z)$. A special case assumes that $H_0(z)$ and $H_1(z)$ are sub-band filters satisfying the mirror filter relationship $H_1(z) = H_0(-z)$ or $H_1(e^{j\omega}) = H_0(e^{j(\omega-\pi)})$. In the z-domain, this assignment results in the QMF filter condition $Y(z) = k\left[H_0^2(z) - H_0^2(-z)\right] X(z) = T(z) X(z)$ where k is a real scale factor introduced by decimation. The filter function $T(z)$ can represent several personalities. The most common persona results in nonideal performance and possible distortion classified as

- ALD = alias distortion;
- AMD = amplitude distortion; and
- PHD = phase distortion.

If a filter is ALD, AMD, and PHD free, the filter possesses the perfect reconstruction (PR) property, and $Y(z) = Kz^{-d}X(z)$, or $y[k] = Kx[k - d]$. That is, the QMF system reconstructs an output that is simply a scaled and delayed version of the input. Since the scale factor and delay are known, the original signal can be reconstructed from $y[k]$.

In the study of FIRs, the virtues of linear phase performance were established on numerous occasions. Suppose it is desired to design an Nth-order QMF FIR that is also linear phase. If N is odd, it can be shown that a null will be placed in the output spectrum at the normalized frequency $\omega = \pi/2$. As a result, an odd order linear phase FIR can remove the aliasing errors from the output, but it cannot have a flat magnitude frequency response. If N is even, then a response that is both linear phase and flat is only produced by a trivial two-coefficient FIR having the form

$$H_0(z) = c_0 z^{-2n_0} + c_1 z^{-2(n_1+1)}; \quad H_1(z) = c_0 z^{-2n_0} - c_1 z^{-2(n_1+1)} \qquad (19.6)$$

for some integer n_0 and n_1. Unfortunately, this filter has little value in practice. Any other even order linear phase choice will result in a distorted filter.

It is known that there does not exist any nontrivial, or physically meaningful, flat response, linear phase QMF filter. Most QMF designs represent some compromise. If the linear phase, or perfect mirror condition (i.e., $H_1(z) = H_0(-z)$), is relaxed, then a magnitude and PHD-less QMF system can be realized. A popular design paradigm is called the perfect reconstruction QMF (PRQMF) method. The output of a PRQMF system is equal to the input with a known delay. The PRQMF design procedure is given by the following:

1. Define a linear phase equirippple FIR $F(z)$ to be a $(2N - 1)$-order half-band FIR having a ripple deviation δ.
2. Classify the zeros of the filter as being interior or exterior to the unit circle. Unfortunately, many of the zeros of $F(z)$ lie on the unit circle and cannot be readily classified as being interior or exterior with respect to the unit circle. Therefore, add $q\delta$ to the center-tap weight of $F(z)$ to form $F_+(z) = F(z) + q\delta$, where $q > 1.0$, but close to unity. This action makes the minimum passband gain $F_+(z)$ bounded by unity. Biasing $F(z)$ in this manner lifts the zeros off the unit circle and forces them to be either interior of exterior.
3. Define an Nth-order FIR $H_0(z)$ and $H_1(z)$ satisfying $F_+(z) = H(z)H(z^{-1})$. Define $H(z)$ in terms of the interior zeros.
4. Let $H_0(z) = H(z)$ and $H_1(z) = (-1)^{N-1} z^{-(N-1)} H(-z^{-1})$.
5. Let $G_0(z) = H_1(-z)$ and $G_1(z) = -H(-z)$.

The result is an all-pass PRQMF system having an input–output transfer function $T(z) = Kz^{-(N-1)}$ where K is a constant.

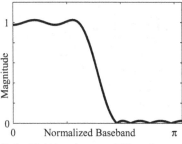 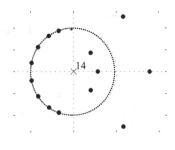

Figure 19.7   Half-band parent FIR and zero distribution of $F_+(z)$.

Example: PRQMF

Design a PRQMF system based on a 15th-order (i.e., order $[2N - 1]$) linear phase half-band filter satisfying $F(z) = -0.02648z^7 + 0.0441z^5 - 0.0934z^3 + 0.3139z^1 + 0.5 + 0.3139z^{-1} - 0.0934z^{-3} + 0.0441z^{-5} - 0.02648z^{-7}$. The magnitude frequency response of $F(z)$ is shown in Figure 19.7. From the half-band FIR, a PRQMF system can be defined by following the stated step-by-step design processes, beginning with the following:

Step 1: $F(z)$ is given and $\delta = 0.0238$.

Step 2: Let $q = 1.01$ and produce $F_+(z)$ such that $F_+(z) = F(z)$ except at $z = 0$ where $F_+(0) = F(0) + q\delta$.

Step 3: The factors of $F_+(z)$ are (up to the precision of the computing routine) the following:

z_i	$|z_i|$	Interior/Exterior
$-0.939 \pm j0.398$	1.02	Exterior
$-0.903 \pm j0.382$	0.98	Interior
$-0.451 \pm j0.907$	1.013	Exterior
$-0.439 \pm j0.884$	0.987	Interior
$-0.394 \pm j0.427$	0.581	Interior
0.561	0.561	Interior
$-1.167 \pm j1.264$	1.72	Exterior
1.782	1.782	Exterior

The location of the 14 zeros of $F_+(z)$ are shown in Figure 19.7. Collecting the zeros residing within the unit circle and multiplying them together, one obtains filters having the following approximate values summarized below:

$$H(z) = 1.0 + 1.34z^{-1} + 0.68z^{-2} - 0.24z^{-3} - 0.34z^{-4} + 0.099z^{-5} + 0.239z^{-6} - 0.17z^{-7}$$
$$= H_0(z),$$

$$H_1(z) = -0.17 - 0.24z^{-1} + 0.099z^{-2} + 0.34z^{-3} - 0.24z^{-4} - 0.68z^{-5} + 1.34z^{-6} - 1.0z^{-7},$$

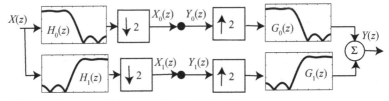

Figure 19.8 $H_0(z)$, $H_1(z)$, $G_0(z)$, and $G_1(z)$ of a QMF filter.

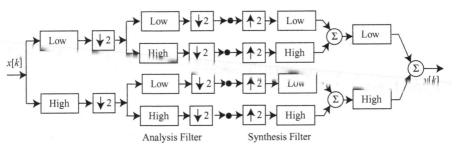

Figure 19.9 Dyadic filter architecture.

$$G_0(z) = -0.17 + 0.24z^{-1} + 0.099z^{-2} - 0.34z^{-3} - 0.24z^{-4} + 0.68z^{-5} + 1.34z^{-6} + 1.0z^{-7},$$
$$G_1(z) = -1.0 + 1.34z^{-1} - 0.68z^{-2} - 0.24z^{-3} + 0.34z^{-4} + 0.099z^{-5} - 0.24z^{-6} - 0.17z^{-7}.$$

Note that individually, the asymmetric filters are also nonlinear phase filters. The frequency response shape and placement of these FIR filters are graphically shown in Figure 19.8. It can be seen that the two channels, before decimation, and after interpolation, have overlapping frequency responses. The condition outlined in Equation 19.5 insures, however, that no aliasing distortion is present in the reconstructed signal.

In practice, the two-channel QMF filter displayed in Figure 19.8, can be used to motivate an $N = 2^n$ channel filter bank having n-levels. The structure of a high-order dyadic filter bank is suggested in Figure 19.9. The analysis filters are $H_0(z)$ (Lo) and $H_1(z)$ (Hi), and the synthesis stage filters are $G_0(z)$ (Lo) and $G_1(z)$ (Hi).

POLYPHASE REPRESENTATION

A two-channel QMF filter can be expressed in polyphase form in terms of $H_0(z) = P_0(z^2) + z^{-1}P_1(z^2)$, where P_0 and P_1 are part of the polyphase representation of $H_0(z)$. It then follows that $H_1(z) = P_0(z^2) - z^{-1}P_1(z^2)$. The resulting QMF is shown in Figure 19.10. The displayed architecture can also be implemented as a modified architecture, also shown in Figure 19.10, that operates at a lower decimated filter clock speed.

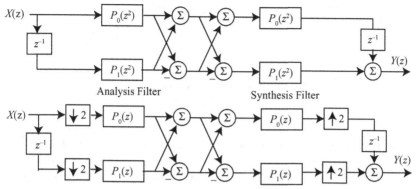

Figure 19.10   Polyphase and modified polyphase representation of a QMF filter.

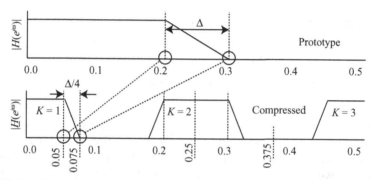

Figure 19.11   Magnitude frequency response of compressed by $M = 4$ FIRs showing the critical frequencies of $\varpi_p = 0.2$ and $\varpi_{a1} = 0.3$ being mapped (compressed) to $\varpi = 0.2/4 = 0.05$ and $\varpi = 0.3/4 = 0.075$. The transition bandwidth is scaled from $\Delta = 0.1$ to $\Delta/M = 0.025$. In addition, multiple copies of the compressed spectra are distributed uniformly along the baseband frequency axis.

FREQUENCY MASKING FILTERS

There are instances when a steep-skirt (i.e., narrow transition band) filter is required. Unfortunately, steep-skirt fixed-sample rate filters are historically very complex and of high order. Such filters can, however, often be designed using multirate techniques based on the frequency masking method. The frequency masking method uses what are called compressed filters. A compressed by M version of a prototype FIR $H(z)$ is $\underline{H}(z) = H(z^M)$ and can be realized by replacing each single clock delay in $H(z)$ with an M sample delay. The compressed filter $\underline{H}(z)$ continues to be clocked at the original sample rate f_s.* Refer to Figure 19.11 and observe how compression scales the frequency axis by a factor $1/M$ and, as a consequence, compresses the FIR's original

---

* A compression filter is not an interpolating FIR. An interpolation FIR operates at an elevated sample rate $f_M = Mf_s$, a compression filter operates at the original rate f_s.

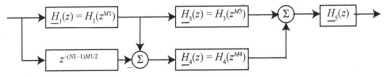

Figure 19.12   Anatomy of a frequency masked (steep skirt) FIR.

transition bandwidth by a like amount. It can also be observed that the act of compression populates the baseband spectrum with multiple copies, or artifacts, of the compressed prototype filter's frequency response. The center frequencies for these artifacts are located at $\varpi_k/\varpi_s = k/M$. It is through the intelligent use of compression that steep-skirt filters can be realized.

The frequency masked FIR architecture, presented in Figure 19.12, consists of the following definable subsystems:

- $\underline{H}_1(z)$, a compressed by M_1 version of an N_1-order FIR $H_1(z)$;
- $\underline{H}_2(z)$, the compressed by M_2 version of the complement of $H_1(z)$;
- $\underline{H}_3(z)$, the compressed by M_3 version of an N_3-order FIR $H_3(z)$;
- $\underline{H}_4(z)$, the compressed by M_4 version of an N_4-order FIR $H_4(z)$;
- $H_5(z)$, an N_5-order FIR.

The compression factor M_1 is chosen in order to map a transition bandwidth of Δ_1 to the final transition bandwidth of $\Delta = \Delta_1/M_1 \ll \Delta_1$. The target filter's low-pass cutoff frequency ϖ_p needs to be made coincident with one of the critical frequencies of a compressed filter $\underline{H}_1(z)$ (e.g., $K_1\varpi_{p1}/M_1$) or the compressed complement filter $\underline{H}_2(z)$ as motivated in Figure 19.12. Notice that the passband trailing edge for the first compressed image (i.e., $K = 0$) is located at $\varpi_{p1}/4 = 0.05$, and for the second image ($K = 1$), at $4\varpi_{p1}/4 = 0.3$. The compressed and compressed complement filters are summarized below:

$$\varpi/\varpi_s = \begin{cases} (K\varpi_s + \varpi_{p1})/\varpi_s M_1; & \underline{H}_1(z) - \text{centric} \\ ((K+1)\varpi(\varpi_s/2) - \omega_p)/\varpi_s M_1; & \underline{H}_2(z) - \text{centric} \end{cases}. \tag{19.7}$$

The stopband critical frequencies can be likewise determined and are a function of M_1 and the original stopband frequency of $H_1(z)$, or those of the complement filter $H_2(z)$. Once the compressed or compressed complement filter's critical frequency is chosen, a need for a housekeeping filter becomes apparent. The compressed artifacts are generated by the compressed prototype and compressed complement prototype filters extending beyond the target filter's passband frequency and need to be eliminated. This is the role of the frequency-masking filters $\underline{H}_3(z)$ and $\underline{H}_4(z)$. The optional last stage shaping FIR $\underline{H}_5(z)$, shown in Figure 19.12, provides a final level of artifact suppression. The rules that codify the design of a frequency masking filter are shown below:

- The component FIR filters $\underline{H}_1(z)$, $\underline{H}_2(z)$, $H_3(z)$, $\underline{H}_4(z)$, and $\underline{H}_5(z)$ should be designed to have their transition bands somewhere in the middle of the baseband range $f \in [0, f_s/2)$. This will ensure that no unusual passband or stopband widths are imposed on the component filters.

- The filter $\underline{H}_1(z)$, or its complement, must have a critical passband frequency ω_p, that maps to the target passband frequency for a compression factor of M_1 and copy index K.

- The design should minimize the solution's transition bandwidth, which can be achieved by minimizing γ where

$$\gamma = \min\left\{\frac{1}{\Delta_{\underline{H}_1(z)}} + \frac{1}{\Delta_{\underline{H}_3(z)}} + \frac{1}{\Delta_{\underline{H}_5(z)}}, \frac{1}{\Delta_{\underline{H}_4(z)}} + \frac{1}{\Delta_{\underline{H}_4(z)}} + \frac{1}{\Delta_{\underline{H}_5(z)}}\right\}. \qquad (19.8)$$

This metric corresponds to the estimated transition bandwidths of the upper and lower paths. The value of γ can be reduced if all the compressed filters have similar transition bandwidths. For the case where the component filters are of differing orders (N_i) and transition bandwidths (Δ_i), the design strategy is to create filters where the values of $N_i\Delta_i$ are similar. As a rule, the highest order FIR section in a frequency-masked system is generally $\underline{H}_1(z)$ (therefore $\underline{H}_2(z)$), followed by $\underline{H}_3(z)$, $\underline{H}_4(z)$, and finally $\underline{H}_5(z)$. This suggests that their individual uncompressed transition bandwidths should appear in the reverse order. For linear phase solutions, the group delay of the upper and lower paths need to be the same. If $N_4 < N_3$, then filter $\underline{H}_4(z)$ will need to be equipped with additional $(N_4 - N_3)/2$ shift register delays in order to equalize the group delays of the upper and lower paths. Finally, as a general rule, the passband deviation of each filter can be chosen to be 25–33% of the target deviation, to account for the degradation (increase) in passband ripple due to cascading.

Example: Compression Filter

Design a steep-skirt FIR low-pass filter having the following specifications:

- passband defined over $f \in [0.0, 0.1]f_s$ (i.e., $\omega_p = 0.1$), with a maximum deviation of -0.175 dB from 0 dB;
- stopband defined over $f \in [0.1025, 0.5]f_s$ (i.e., $\omega_a = 0.1025$), with a gain of -40 dB, or less; and
- a transition bandwidth of $(0.1025 - 0.1)f_s = (0.0025)f_s$.

It is worth noting that satisfying the specifications with a single speed filter would require a linear phase equiripple filter having an order in excess of 700. This is, in most instances, unacceptable. The design of a steep-skirt filter begins with a definition of the prototype $\underline{H}_1(z)$ in terms of the critical design parameters $(\omega_{p1}, \omega_{a1}, \Delta_1)$ and compression ratio (M_1) and replication constant (K_1). Since the target normalized transition bandwidth satisfies $0.0025 = \Delta_1/M_1$, a list of acceptable Δ_1 and M_1 pairs can be assembled using a direct computer search. A reasonable, but by no means unique choice is $M_1 = 17$ resulting in $\Delta_1 = 0.0428$. Next, for $M_1 = 17$, the targeted passband cutoff frequency needs to be expressed in terms of the compression filter parameters ω_{p1} or ω_{a1}, and K_1 (these parameters also apply to the compressed complement filter as well.) Again a direct computer search can be used to sort out

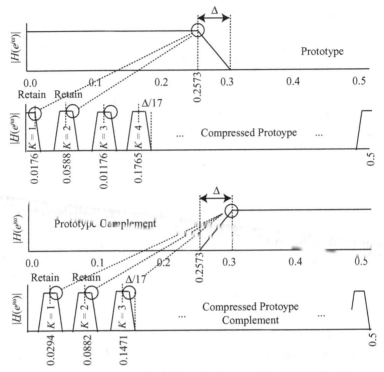

Figure 19.13   Frequency-masked design example for $\varpi_p = 0.1$, $M = 17$, $K = 2$, $\varpi_{p1} = 0.2573$, and $\varpi_{a1} = 0.3$. The final design retains two copies ($K = 2$) of the compressed prototype spectrum and two copies ($K = 2$) of the compressed complement response.

the parametric options as illustrated in Figure 19.11. Figure 19.13 shows the outcome if $\varpi_p = 0.1$, $\varpi_{p1} = 0.2573$, $\varpi_{a1} = 0.3$, $K = 2$, and $M = 17$. That is, the target passband frequency is obtained by compressing the complement filter $\underline{H_2}(z)$ and $K = 2$.

Finally, the passband gains of the component filters need to be specified. Suppose the minimum passband gain for the upper path is $G_{upper} = G_1 G_3 G_5$ and $G_{lower} = G_2 G_4 G_5$ for the lower path. Assume, for the purpose of discussion, that all individual gains are comparable such that $G_{upper} \sim G_1^3$ and $G_{lower} \sim G_1^3$. Since the specified minimum passband gain is on the order of -0.175 dB, it follows that $G_{upper} = G_{lower} = 0.98$ or the gain deviation is $(1 - 0.98) = 0.02 \sim -34$ dB. For design consistency, let the passband deviation of all the filters (i.e., $G_1, \ldots, G_5$) be essentially the same. Then if $G_{upper} = G_{lower} = 0.98 = G_1^3$, or $G_1 = 0.9933$ resulting in a passband deviation on the order of $1 - 0.9933 = 0.0067 \sim -43$ dB. The filters $\underline{H_3}(z)$ and/or $\underline{H_4}(z)$ will spawn spectral artifacts that are outside the final solution's passband, which are suppressed by $\underline{H_5}(z)$. The locally minimum stopband attenuation is essentially set by the stopband attenuation of filter $\underline{H_5}(z)$.

The frequency-masking filter process is summarized in Table 19.1. All component filters are equiripple FIRs with critical frequencies ϖ_{pi} and ϖ_{ai} as described in Table 19.1. The behavior of the filter $\underline{H_2}(z)$ is established by $\underline{H_1}(z)$. The filters are

TABLE 19.1.   Frequency Masked Filter

Item	$H_1(z)$	$H_2(z)$	$H_3(z)$	$H_4(z)$	$H_5(z)$
Passband edge (ϖpi)	0.257200	0.300000	0.229412	0.300000	0.100000
Stopband edge (ϖai)	0.300000	0.257200	0.307500	0.398382	0.200599
Passband ripple (δ_{pi})	−43 dB	−56 dB	−45 dB	−58 dB	−42 dB
Stopband ripple (δ_{ai})	−56 dB	−43 dB	−58 dB	−71 dB	−42 dB
Filter order N_i	63	63	37	37	23
$N_i\Delta_i$	2.8676	2.8676	2.8897	3.64	2.438
Transition bandwidth Δ_i	0.0428	0.0428	0.0781	0.0984	0.1006
Compression factor M_i	17	17	3	3	1

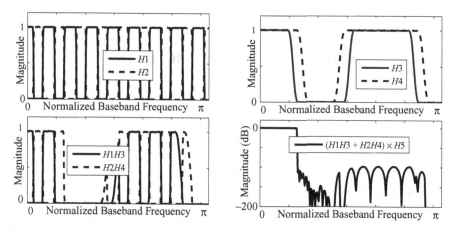

Figure 19.14   Frequency masked filter component and localized response. The lower left panel displays the output response.

generally designed to have a passband deviation on the order of −40 dB and a value of $N_i\Delta_i \sim 2.8$. The exception is the 37th-order $H_4(z)$ (same order as $H_3(z)$), where $N_4\Delta_4 \sim 3.6$. A 29th-order $H_4(z)$ could have been used resulting in an $N_4\Delta_4 \sim 2.85$ if eight addition delays (four predelays, four postdelays) are added to equalize group delays. Choosing a 37th-order FIR over a 29th-order FIR will simply result in the lower path having a slightly different gain deviation. The filter $H_5(z)$ is chosen to suppress the high-frequency anomalies shown in Figure 19.14

Figure 19.14 shows the spectral response of the complete 160th-order solution (Note: $H_2(z)$ is assumed to be implemented as a delay-enabled complement FIR requiring no additional coefficient multipliers). The design is based on a compressed critical frequency obtained from $H_2(z)$ ($\varpi_{a1} = 0.3$ is compressed to $\varpi_p = 0.1$). It can also be seen that the filters $H_3(z)$ and $H_4(z)$ pass the first $K_1 = 2$ copies of $H_1(z)$ and $H_2(z)$, respectively. The estimated transition bandwidths of the upper and lower paths are

$$\frac{1}{\Delta_{\text{upper}}} = \frac{1}{\Delta_{\hat{H}_1}} + \frac{1}{\Delta_{\hat{H}_3}} + \frac{1}{\Delta_{\hat{H}_5}} = \frac{1}{0.0428/17} + \frac{1}{0.0781/3} + \frac{1}{0.1006} = \frac{1}{0.0216},$$

$$\frac{1}{\Delta_{\text{lower}}} = \frac{1}{\Delta_{\hat{H}_2}} + \frac{1}{\Delta_{\hat{H}_4}} + \frac{1}{\Delta_{\hat{H}_5}} = \frac{1}{0.0428/17} + \frac{1}{0.0984/3} + \frac{1}{0.1006} = \frac{1}{0.0230},$$

which results in a value of $\Delta_{\text{steep-skirt}} = 0.0023 < 0.0025$. The resulting steep-skirt linear phase FIR is analyzed in Figure 19.14 and exhibits a

- passband edge: $0.1f_s$,
- stopband edge: $0.1025f_s$,
- passband ripple: <0.1 dB, and
- stopband ripple: >42 dB,

which are seen to meet or exceed the design specifications.

CASCADED INTEGRATOR-COMB (CIC) FILTER

The motivation to develop wireless systems with high digital content has sometimes been based on valid arguments, at other times, false assumptions. While consumers have seen significant changes in packaging and functionality of wireless instruments, the underlying physical infrastructure has evolved slowly. Consider, as an example, the wireless transceiver shown in Figure 19.15. The engineering desire is to "drive" the digital portion of the transceiver as close to the antenna as possible, or practical. The digital portion of the system is only incrementally expanding. This slow growth is the result of the fact that mobile wireless devices are power sensitive, which precludes using ultra high-speed digital logic and analog-to-digital converters (ADCs). While it is possible to digitize RF or near RF signals in the GHz+ range, such technologies would have a power dissipation budget that would preclude its use in mobile wireless applications. Nevertheless, there remains a constant problem of locating where to initiate the requisite analog-to-digital conversion. Assuming that 300 MHz

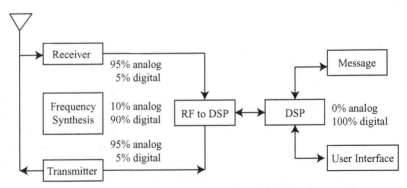

Figure 19.15 Anatomy of a wireless transceiver. DSP, digital signal processing.

may represent an upper bound for low-power ADC operation, it is apparent that 1–5 GHz (or beyond) wireless signals will still continue to require embedded analog processing to reduce a signal's bandwidth to a value consistent with a low-power ADC. Once a signal is digitized, it can be processed using various algorithms. The front-end baseband bandwidth must be equal to, or in excess of the signal bandwidth defined by the modulation scheme. For narrow-band communication applications, the channel bandwidth is much lower than the first ADC rate. For broadband applications, the signal bandwidth is assumed to be 1/10th (or more) of f_s.

The conversion of a digitized signal (e.g., 300 MHz) to baseband (e.g., 300 kHz) is the role of a digital down converter (DDC), or channelizer. The basic problem in implementing this sample rate conversion is, again, the speed limitations of digital signal processing elements. If a linear phase FIR is used to select a narrow-band of frequencies from a broadband spectrum, the resulting FIR order can become excessively high, requiring a large number of multiply-accumulate (MAC) calls per filter cycle. Implementing these filters in real time (e.g., 300 MHz) would place an excessive power demand of the design. A properly designed channelizer, however, can mitigate this problem. The preferred channelizer architecture is called a CIC or Hogenauer filter.*

A complete down converter system is suggested in Figure 19.16. The CIC architecture has a definite advantage over conventional digital filters in that it is multiplier-less (i.e., MAC free). The CIC solution consists of a direct digital synthesizer (local carrier generator) and a digital mixer that heterodynes a desired subband down to DC. A CIC filter section includes a collection of comb and integrator filters separated by a decimation by R circuit. Each integrator has a transfer function given by $H_I(z) = 1/(1 - z^{-1})$ and is clocked at the ADC rate of f_s. It can be noted that the integrator requires only a shift register and adder (no multiplier) to implement. Each integrator has a pole located at $z = 1$. Cascading N integrators together results in a partial transfer function of $H_I(z) = 1/(1 - z^{-1})^N$, which contributes N poles at $z = 1$ to the design. Following the integrators are N comb filters. Each comb filter is clocked at a decimate rate of $f_{comb} = f_s/R$. The individual transfer function for each comb filter is therefore $H_C(z) = (1 - z^{-R})$, which contributes R zeros distributed

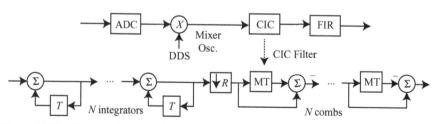

Figure 19.16   Down converter (channelizer) architecture. Conventional signal processing heterodyne solution involving a channelizer (a.k.a., digital down converter). The Nth-order channelizer consists of N integrators and N delay M comb filter, separated by a decimate-by-R circuit. Osc., oscillator; DDS, direct digital synthesizer; MT, as is.

---

* MATLAB's mfilt.cicdecim command in the Filter Design Toolbox provides some CIC support.

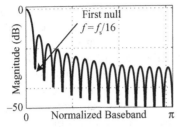 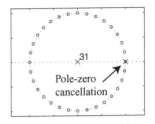

Figure 19.17 Baseband CIC filter section ($R = 32$).

uniformly about the unit circle on multiples of π/R radian centers, including one at $z = 1$. Cascaded together, the N comb filter section results in $H_C(z) = (1 - z^{-R})^N$. The result is called an Nth-order CIC filter displayed in Figure 19.17, and is characterized by a transfer function $H_{CIC}(z) = (1 - z^{-R})^N/(1 - z^{-1})^N$. The decimation rate establishes the output baseband range for 0 to the first null at f_s/P. The Nth order CIC filter possesses N-poles, located at DC (i.e., $z = 1.0$), and R zeros of multiplicity N located on the unit circle at $z = e^{j2\pi k/R}$, $k \in [0, R)$ including N zeros at $z = 1$. The N zeros located at $z = 1$ cancel an equal number of zeros at the same location (see Fig. 19.17) resulting in a filter having a high DC gain (i.e., low-pass). The pole-zero cancellation is ensured because all the filter coefficients are exact at $z = 1$. Furthermore, since all the CIC filter coefficients are unit-valued, they require no general-purpose processor or embedded MAC to realize. Because of this, CIC filters are often used in high-speed communication and instrument applications at the physical layer.

BIBLIOGRAPHY

Adali, J., and Hayken, S. *Adaptive Signal Processing: Next Generation Solutions.* Wiley, 2010.

Antoniou, A. *Digital Filters: Analysis, Design, and Applications,* 2nd edition. McGraw-Hill, 1993.

Bose, T. *Digital Signal and Image Processing.* Wiley, 2003.

Cavicchi, T. *Digital Signal Processing.* Wiley, 2000.

Chaparro, L. *Signals and Systems Using MATLAB.* Elsevier, 2011.

Chassaing, R. *Digital Signal Processing with the C6713 and C6416 DSK.* Wiley, 2004.

Chassaing, R., and Reay, D. *Digital Signal Processing and Applications with the TMS320C6713 DSK.* Wiley-IEEE, 2008.

Chen, C.-T. *Digital Signal Processing.* Oxford University Press, 2001.

Clark, C. *LabVIEW Digital Signal Processing.* McGraw-Hill, 2005.

Corinthios, M. *Signals, Systems, Transforms, and Digital Signal Processing with MATLAB.* CRC Press, 2009.

Cristi, R. *Modern Digital Signal Processing.* Thompson, 2004.

Grover, D., and Diller, J. *Digital Signal Processing.* Prentice Hall, 1999.

Hayes, M. *Schaum's Outline of Digital Signal Processing.* Schaum, 1999.

Heck, B., and Kamen, E. *Fundamentals of Signals and Systems Using the Web and MATLAB.* Prentice Hall, 2007.

Ifwachor, E., and Jervis, B. *Digital Signal Processing,* 2nd edition. Addison Wesley, 2001.

Ingle, V., and Proakis, J. *Digital Signal Processing Using MATLAB.* Cengage Learning, 2007.

Jackson, L. *Digital Filters and Signal Processing.* Kluwer, 1989.

Kehtarnavaz, N., and Kim, N. *Digital Signal Processing System Design Using LabVIEW.* Newnes, 2005.

Kuo, S., and Gan, W.-S. *Digital Signal Processing.* Prentice Hall, 2004.

Lutovac, M., Tosic, D., and Evans, B. *Filter Design for Signal Processing, Using MATLAB and Mathematica.* Prentice Hall, 2001.

Lyons, R. *Understanding Digital Signal Processing.* Prentice Hall, 2004.

Madisetti, V. *The Digital Signal Processing Handbook,* 2nd edition. CRC Press, 2009.

McClellan, J., Schafer, R., and Yoder, M. *Signal Processing First.* Prentice Hall, 1998.

Digital Filters: Principles and Applications with MATLAB, First Edition. Fred J. Taylor.
© 2012 by the Institute of Electrical and Electronics Engineers, Inc.
Published 2012 by John Wiley & Sons, Inc.

Mersereau, R., and Smith, M. *Digital Filtering*. Wiley, 1993.

Mitra, S. *Digital Signal Processing Laboratory Using MATLAB*. McGraw-Hill, 1999.

Mitra, S. *Digital Signal Processing: A Computer-Based Approach*, 4th edition. McGraw-Hill, 2011.

Moon, T., and Stirling, W. *Mathematical Methods and Algorithms for Signal Processing*. Prentice Hall, 2000.

Oppenheim, A., and Schafer, R. *Digital Signal Processing*, 2nd edition. Prentice Hall, 1999.

Oppenheim, A., and Schafer, R. *Discrete-Time Signal Processing*, 3rd edition. Prentice Hall, 2010.

Orfanidis, S. *Signal Processing*. Prentice Hall, 1996.

Parhi, K. *VLSI Ditial Signal Processing and Implementation*. Wiley, 1995.

Parks, T., and Burrus, C. *Digital Filter Design*. Wiley, 1987.

Phillips, C., Parr, J., and Riskin, E. *Signals, System, and Transforms*, Prentice Hall, 2003.

Poral, B. *A Course in Digital Signal Processing*. Wiley, 1997.

Proakis, J., and Manaolakis, D. *Digital Signal Processing*, 4th edition. Prentice Hall, 2007.

Rao, P. *Signals and Systems*. McGraw-Hill, 2008.

Saben, W.E. *Discrete-Time Signal Analysis and Design*. Wiley, 2008.

Schilling, R., and Harris, S. *Fundamentals of Digital Signal Processing Using MATLAB*. Thomson, 2005.

Schlichtharle, D. *Digital Filters*. Springer, 2000.

Schuler, C., and Chugani, M. *Digital Signal Processing*. McGraw-Hill, 2004.

Searns, S., and Hush, D. *Digital Signal Processing with Examples in MATLAB*. CRC Press, 2011.

Singh, A., and Srinivansan, S. *Digital Signal Processing*. Thompson, 2004.

Smith, S. The scientists and engineer's guide to digital signal processing. www.DSPguide.com.

Taylor, F. *Digital Filter Design Handbook*. Marcel Dekker, 1983.

Taylor, F., and Mellott, J. *Hands-On Digital Signal Processing*. McGraw-Hill, 1995.

Wang, B.-C. *Digital Signal Processing Techniques and Application in Radar Image Processing*. Wiley, 2008.

Widrow, B., and Kollar, I. *Quantization Noise*. Cambridge, 2008.

Williams, A., and Taylor, F. *Electronic Filter Design Handbook*, 4th edition. McGraw-Hill, 2006.

Yates, W. *Digital Signal Processing*. CSP, 1989.

Zelniker, G., and Taylor, F. *Advanced Digital Signal Processing*. Dekker, 1994.

APPENDIX

MATLAB

MATLAB contains many signal processing and digital filter design functions. Some of the more popular and germane programs and utilities are listed below and found in

> MATLAB,
>
> MATLAB Signal Processing Toolbox,
>
> MATLAB Filter Design Toolbox,
>
> MATLAB Control Toolbox, and/or
>
> MATLAB Communication Toolbox.

Signal Processing Functions in MATLAB

conv	Convolution and polynomial multiplication
fft	Fast Fourier transform
fftshift	Shift zero-frequency component to center of spectrum
ifft	Inverse discrete Fourier transform
unwrap	Correct phase angles to produce smoother phase plots

Digital Filters

Discrete-Time Filters

dfilt.cascade	Cascade of discrete-time filters
dfilt.delay	Delay filter
dfilt.df1	Discrete-time, direct-form I filter
dfilt.df1sos	Discrete-time, second-order section, direct-form I filter
dfilt.df1t	Discrete-time, direct-form I transposed filter
dfilt.df1tsos	Discrete-time, second-order section, direct-form I transposed filter

---

Digital Filters: Principles and Applications with MATLAB, First Edition. Fred J. Taylor.
© 2012 by the Institute of Electrical and Electronics Engineers, Inc.
Published 2012 by John Wiley & Sons, Inc.

dfilt.df2sos	Discrete-time, second-order section, direct-form II filter
dfilt.df2t	Discrete-time, direct-form II transposed filter
dfilt.df2tsos	Discrete-time, second-order section, direct-form II transposed filter
dfilt.dfasymfir	Discrete-time, direct-form antisymmetric finite impulse response (FIR) filter
dfilt.dffir	Discrete-time, direct-form, FIR filter
dfilt.dffirt	Discrete-time, direct-form FIR transposed filter
dfilt.dfsymfir	Discrete-time, direct-form symmetric FIR filter
dfilt.fftfir	Discrete-time, overlap-add, FIR filter
dfilt.latticeallpass	Discrete-time, lattice, all-pass filter
dfilt.latticear	Discrete-time, lattice, autoregressive filter
dfilt.latticearma	Discrete-time, lattice, autoregressive, moving-average filter
dfilt.latticemamax	Discrete-time, lattice, moving-average filter
dfilt.latticemamin	Discrete-time, lattice, moving-average filter
dfilt.parallel	Discrete-time, parallel structure filter
dfilt.scalar	Discrete-time, scalar filter
dfilt.statespace	Discrete-time, state-space filter

FIR Filter Design

cfirpm	Complex and nonlinear phase equiripple FIR filter design
fir1	Window-based FIR filter design
fir2	Frequency sampling-based FIR filter design
fircls	Constrained least square, FIR multiband filter design
fircls1	Constrained least square, low-, high-pass, linear phase, FIR filter design
firls	Least square linear phase FIR filter design
firpm	Parks–McClellan optimal FIR filter design
firpmord	Parks–McClellan optimal FIR filter order estimation
intfilt	Interpolation FIR filter design
kaiserord	Kaiser window FIR filter design estimation parameters sgolay Savitzky–Golay filter design

Infinite Impulse Response (IIR) Digital Filter Design

butter	Butterworth filter design
cheby1	Chebyshev Type I filter design (passband ripple)
cheby2	Chebyshev Type II filter design (stopband ripple)
ellip	Elliptic filter design
maxflat	Generalized digital Butterworth filter design
yulewalk	Recursive digital filter design

IIR Filter Order Estimation

buttord	Butterworth filter order and cutoff frequency
cheb1ord	Chebyshev Type I filter order
cheb2ord	Chebyshev Type II filter order
ellipord	Minimum order for elliptic filters

Filter Analysis

abs	Absolute value (magnitude)
angle	Phase angle
freqz	Frequency response of digital filter
fvtool	Open Filter Visualization Tool
grpdelay	Average filter delay (group delay)
impz	Impulse response of digital filter
phasedelay	Phase delay of digital filter
phasez	Phase response of digital filter
stepz	Step response of digital filter
zerophase	Zero-phase response of digital filter
zplane	Zero-pole plot

Filter Discretization

bilinear	Bilinear transformation
impinvar	Impulse invariance transformation

Linear Systems

latc2tf	Convert lattice filter parameters to transfer function form
residuez	z-Transform partial-fraction expansion
sos2ss	Convert digital filter second-order section parameters to state-space form
sos2tf	Convert digital filter second-order section data to transfer function form
sos2zp	Convert digital filter second-order section parameters to zero-pole-gain form
ss2sos	Convert digital filter state-space parameters to second-order sections form
ss2tf	Convert state-space filter parameters to transfer function form
ss2zp	Convert state-space filter parameters to zero-pole-gain form
tf2latc	Convert transfer function filter parameters to lattice filter form
tf2sos	Convert digital filter transfer function data to second-order sections form
tf2ss	Convert transfer function filter parameters to state-space form
tf2zp	Convert transfer function filter parameters to zero-pole-gain form
tf2zpk	Convert transfer function filter parameters to zero-pole-gain form
zp2sos	Convert zero-pole-gain filter parameters to second-order sections form
zp2ss	Convert zero-pole-gain filter parameters to state-space form
zp2tf	Convert zero-pole-gain filter parameters to transfer function form

Windows

barthannwin	Modified Bartlett–Hann window
bartlett	Bartlett window
blackman	Blackman window
blackmanharris	Minimum four-term Blackman–Harris window
bohmanwin	Bohman window
chebwin	Chebyshev window
dpss	Discrete prolate spheroidal (Slepian) sequences
dpssclear	Remove discrete prolate spheroidal sequences from database
dpssdir	Discrete prolate spheroidal sequences database directory
dpssload	Load discrete prolate spheroidal sequences from database
flattopwin	Flat top weighted window

gausswin	Gaussian window
hamming	Hamming window
hann	Hann (Hanning) window
kaiser	Kaiser window
nuttallwin	Nuttall-defined minimum four-term Blackman–Harris window
parzenwin	Parzen (de la Vallée-Poussin) window
rectwin	Rectangular window
taylorwin	Taylor window
triang	Triangular window
tukeywin	Tukey (tapered cosine) window
window	Window function gateway
wvtool	Open Window Visualization Tool

Multirate Signal Processing

decimate	Decimation—decrease sampling rate
downsample	Decrease sampling rate by integer factor
interp	Interpolation—increase sampling rate by integer factor
resample	Change sampling rate by rational factor
upfirdn	Upsample, apply FIR filter, and downsample
upsample	Increase sampling rate by integer factor

Waveform Generation

chirp	Swept-frequency cosine
diric	Dirichlet or periodic sinc function
gauspuls	Gaussian-modulated sinusoidal pulse
gmonopuls	Gaussian monopulse
pulstran	Pulse train
rectpuls	Sampled aperiodic rectangle
sawtooth	Sawtooth or triangle wave
sinc	Sinc
square	Square wave
tripuls	Sampled aperiodic triangle
vco	Voltage-controlled oscillator

Graphical User Interfaces (GUIs)

fdatool	Open Filter Design and Analysis Tool
fvtool	Open Filter Visualization Tool
sptool	Open interactive digital signal processing tool
wintool	Open Window Design and Analysis Tool
wvtool	Open Window Visualization Tool

GLOSSARY

Aliasing   Aliasing is an effect that causes a signal, upon sampling, to impersonate another signal.

Amplitude Modulation (AM)   A modulation scheme that carries information on the amplitude envelope of a modulated signal.

Analog Prototype Filter   Classic analog Butterworth, Chebyshev, or elliptic low-pass filter models having a 1 rad/s passband.

Analog-to-Digital (A/D) Converter (ADC)   A device that converts analog signals into sample values having digital precision.

Anti-Aliasing Filter   An analog low-pass filter placed in front of an analog-to-digital converter (ADC) to eliminate or suppress distorting aliasing effects.

Application-Specific Integrated Circuits (ASICs)   Semiconductor devices designed to perform a fixed versus general-purpose function.

Architecture   Architecture refers to the interconnection of fundamental building block elements to define a physical digital signal processing (DSP) solution.

Arithmetic Error   Arithmetic errors occur when performing arithmetic operations using data of finite precession.

Arithmetic Logic Unit (ALU)   A part of a digital processor that contains the arithmetic and logical processing agents.

Autocorrelation   A measure of the similarity of a signal with itself over a range of delays.

Autoregressive Moving Average (ARMA)   A process that can produce an approximate model of a linear system based on direct measurements of a system's frequency response.

Band-Limited Signal   Associated with signals having a finite maximum frequency.

Band-Pass Filter   A filter that passes only a range of frequencies bounded away from 0 Hz (DC) and the Nyquist frequency.

Band-Stop Filter   A filter that suppresses a single range of frequencies bounded away from 0 Hz (DC) and the Nyquist frequency.

Bandwidth   A contiguous range of frequencies that contain signal power or energy.

Baseband   Contiguous frequency range between 0 Hz (DC) and the Nyquist frequency.

Benchmark   A device or process that measure the performance or attributes of a system or subsystem.

Bilinear z-Transform   A type of z-transform well suited for use with frequency-selective infinite impulse response (IIR) filter designs.

Digital Filters: Principles and Applications with MATLAB, First Edition. Fred J. Taylor.
© 2012 by the Institute of Electrical and Electronics Engineers, Inc.
Published 2012 by John Wiley & Sons, Inc.

Biquad A common first- and second-order filter building block architecture.

Bounded-Input Bounded-Output (BIBO) Stable Property of systems producing only bounded outcomes for any arbitrary bounded input.

Butterfly A complex arithmetic operation associated with the production of a fast Fourier transform (FFT).

Canonic Architecture A filter architecture that implements an Nth-order filter with at most N shift registers.

Canonic Signed Digit (CSD) An arithmetic code that is dense in zero digits.

Cascade Architecture A serial connection of low-order filter sections to create a higher-order filter.

Cascaded Integrator-Comb (CIC) Filter Multiplier-less low-pass filter also known as a Hogeneaur filter.

Causal Signal Causal signal has a finite starting time (time of origin).

Chirp z-Transform A method of computing z-transforms using convolution methods.

Circular Convolution A process of convolving two periodic signals with common period.

Classic Infinite Impulse Response (IIR) IIRs based on analog Butterworth, Chebyshev I and II, or elliptic filter models.

Coefficient Round-Off Error An error that occurs when real coefficients are quantized into finite precision data words.

Comb Filter Multiplier-less filter architecture having multiple nulls located at periodically spaced frequencies.

Companding A logarithmic amplitude suppression scheme used to compress the dynamic range of a signal.

Complement Filter A filter that when additively combined with its parent filter, produces an all-pass filter.

Complex Instruction Set (CISC) A digital processor technology having a numerically large instruction set.

Compression A process by which the size of a sampled signal can be reduced without significantly reducing signal information.

Continuous Time Signals or systems that are continuously resolved in both time and amplitude.

Convolution A mathematical process that defines linear filtering.

Correlation A measure of the similarity of two signals over a range of delays.

Critical Sampling Sampling at the Nyquist sample rate $f_s/2$.

Data Flow Architecture Multiprocessing architectures where individual processing elements perform multiple instructions on many pieces of data.

Decimation The act of discarding samples from a time series to produce a sparser time series at a lower sample rate.

Digital Filter A device that is capable of altering a digital signal's magnitude and/or phase response using digital technology.

Digital Signal Processing (DSP) The art and science of creating, modifying, and manipulating signal attributes using digital technology.

Digital Signal Processing Microprocessor (DSP µp) A function-specific processor used to realize DSP solutions.

Digital Signal A digital signal is discretely resolved or quantized in both time and amplitude.

Digital-to-Analog Converter (DAC) A device that converts digital signal into discrete-time or analog signal.

Direct Architecture Filter architecture appearing in canonic (direct II) and noncanonic (direct I) form.

Discrete-Cosine Transform (DCT) A time-frequency transform based on a real cosine expansion.

Discrete Fourier Transform (DFT) An algorithm that maps a discrete-time time-series into the discrete frequency domain.

Discrete-Time Fourier Series (DTFS) The discrete-time Fourier transform of discrete-time periodic signals.

Discrete-Time Fourier Transform (DTFT) The discrete-time Fourier transform of a discrete time series.

Discrete-Time Sample A sample having finite resolution in time and continuous resolution in amplitude.

Discrete-Time Signal A time series consisting of contiguous discrete-time samples.

Distributed Arithmetic A special fixed-point filter architecture that replaces a general purpose multiply-accumulator with memory table lookups.

Dithering A technique of adding a small amount of noise to a signal to reduce the overall statistical error of a reconstructed signal.

Dynamic Range The attainable (unconstrained) range of a signal.

Echo Canceller A filter designed to remove reflected (returned) signals.

Equalization A process that compensates for the nonlinear effects of a channel or system.

Equiripple Finite Impulse Response (FIR) A type of FIR satisfying a minimax ripple error criterion.

Farrow Filter A multirate filter possessing fractional delays.

Fast Fourier Transform (FFT) A fast implementation of a discrete Fourier transform (DFT).

Filter A device that alters the attributes of input signal (e.g., magnitude and phase).

Filter Bank A collection of individual filters designed to collectively achieve a desired frequency-domain effect.

Finite Impulse Response (FIR) Filter A nonrecursive or transversal filter that has no feedback paths.

Finite Word Length Effects Finite errors and uncertainties introduced by finite precision arithmetic.

First-Order Hold A linear interpolation technique used to convert piecewise constant signals into piecewise linear signal.

Fixed Point Numbering or arithmetic system of finite precision and dynamic range.

Fixed-Point Processor A processor used to implement fixed-point arithmetic.

Floating Point A number scheme that codes data with high precision and dynamic range.

Floating-Point Operations Per Second (FLOPS) A measurement of performance of capabilities of a floating-point processor.

Floating-Point Processor A processor used to implement floating-point arithmetic operations.

Frequency First time derivative of phase or, in a popular sense, the number of cycles per unit of time.

Frequency Division Multiplexing (FDM) A process that divides a channel into nonoverlapping frequency bands.

Frequency Domain A domain in which signal and systems are defined in terms of their frequency attributes.

Frequency Masking Filter A multirate finite impulse response (FIR) filter strategy used to synthesize "steep-skirt" FIRs.

Frequency Sampling Filter A filter implemented as a bank of frequency-selective filters.

Gain The amplification ability of a system.

Group Delay The frequency-dependent signal propagation delay introduced by a digital filter.

Half-Band Filter A filter having magnitude frequency response symmetry about $f_s/4$ resulting in a low complexity design.

Harvard Architecture A processor architecture that uses separate busses for program, memory, and input/output (I/O).

Heaviside Expansion A process by which a transfer function can be represented as the additive collection of low-order terms, sometimes called a partial fraction expansion.

High-Pass Filter A filter that passes only high frequencies.

Hilbert Filter A filter having quadrature phase behavior in the frequency domain.

Hogeneaur Finite Impulse Response (FIR) See cascaded integrator-comb filter.

Homogeneous Solution The solution to a difference or differential equation due only to the system's initial conditions.

Image Processing The process of enhancing or extracting information from an image.

Impulse Invariant A property that ensures the sample value of a discrete-time impulse response agrees with those of the continuous-time parent at the sample instances.

Impulse Response The response of a filter to an input impulse signal.

Impulse Sampler An ideal device that can instantaneously sample an analog signal.

Infinite Impulse Response (IIR) Filter Also called a recursive filter, identifies a class of filters having feedback.

Inhomogeneous Solution Solution to a difference or differential equation due to the system's external forcing function.

Integrated Circuit (IC) A semiconductor device or chip.

Interpolating Filter An analog filter that converts a discrete-time signal into an analog signal.

Interpolation The act of increasing the effective sample rate of a signal or synthesizing signal values in between adjacent samples.

Lattice/Ladder Architecture A special type of finite impulse response (FIR) and infinite impulse response (IIR) filter architecture.

L-Band Finite Impulse Response (FIR) Filter exhibiting specific symmetry conditions in the frequency domain.

Least Mean Square (LMS) A filter design optimization criteria.

Limit Cycling   The result of unwanted oscillations due to data rounding, truncation, or saturation.

Linear Phase   A filter property that requires that the filter's propagation delay is constant for all frequencies.

Linear Shift Invariant (LSI) Filter   A linear filter with constant coefficients.

Low-Pass Filter   A filter that passes only low frequencies.

Magnitude Frequency Response   The absolute value of a signal's or system's frequency response.

Microprocessor   Compact full-function general-purpose computers.

Million Instructions Per Second (MIPS)   A measurement of the performance, or capacity, of a digital processor.

Minimum Phase Finite Impulse Response (FIR)   An FIR having zeros only on or interior to the unit circle in the z-plane.

Mirror Filter   A filter having a reflected magnitude frequency response of a parent filter.

Mixed-Signal Device   A system containing a collection of analog and digital signals.

Modular Arithmetic   A type of algebra based on algebraically manipulating remainders.

Modulation   The modification of a signal process using multiplicative processes.

Moving Average Filter   A low complexity low-pass finite impulse response (FIR) filter.

Multiply-Accumulate (MAC)   Operations of the form $S \leftarrow AX + Y$ (see SAXPY).

Multiprocessing   A paradigm in which the computational process is spread across several processors to improve the performance of the system.

Multirate   A condition in which a system contains two or more differing sample rates.

Multitasking   A division of a computational process across several tasks and facilitators.

Noise Gain   The amplification of errors due to internal finite word length effects measured at a system's output.

Noise Power   The power amplification of errors due to internal finite word length effects measured at a system's output.

Normal Architecture   An infinite impulse response (IIR) filter architecture exhibiting state space symmetry.

Nyquist Filter   Finite impulse responses (FIRs) exhibiting special symmetry conditions.

Nyquist Frequency   One-half the sample frequency $f_s/2$.

Nyquist Sample Rate   The minimum sample rate at which an analog signal that enables signal reconstructed without aliasing.

Oversampling   A sampling above the Nyquist sample rate.

Overflow Saturation   Refers to a class of error that occurs when a variable exceeds the dynamic range limitation of a system.

Parallel Architecture   An architecture in which low-order systems are added together to create a higher-order system.

Parallel Processing   Task speedup using concurrent processing.

Periodogram   Power spectrum approximation technique.

Phase Response   Frequency-dependent phase angle of a signal's or system's frequency response.

Pipelining   A technique used to reduce a task into a serial set of smaller tasks that can be individually processed and recombined.

Pole   An artifact (singularity) generated by a feedback path of a filter.

Polyphase Representation   A mathematical procedure of representing decimated and interpolated signals and systems.

Prony Filter   A filter synthesis technique that can be used to model a physical filter's impulse response.

Quadrature Mirror Filter (QMF)   A filter bank capable of reconstructing a signal from its component parts.

Quantization   A process by which a real variable is converted into a digital word of finite precision.

Quantization Error   The error associated with representing a real number with a quantized number.

Recursive Filter   See infinite impulse response (IIR).

Reduced Adder Graph   A technique used to represent a number with a spare set of digits.

Reduced Instruction Set Computer (RISC)   A computer architecture having a small instruction set.

Region of Convergence (ROC)   The region of the z-plane in which the z-transform is guaranteed to exist.

Resolution   The measure of accuracy of an analog-to-digital (ADC) or digital-to-analog (DAC) converter.

Sallen–Key Filter   An analog filter based on an operational amplifier, typically used as an anti-aliasing filter.

Sample and Hold (S&H)   An analog device that captures a sample of an analog signal and holds that value for a simple period.

Sample Frequency   See sample rate.

Sample Rate   The rate at which an analog signal is sampled in samples per second (Sa/s).

Sampling   The process of converting of a continuous time analog signal into a discrete-time time series.

Sampling Theorem   Establishes the conditions under which alias-free signal reconstruction can take place.

Savitzky–Golay Finite Impulse Response (FIR)   A filter form used to perform interpolation using polynomials.

SAXPY   A fundamental digital signal processing (DSP) operation of the form $S \leftarrow AX + Y$ (see multiply-accumulate [MAC]).

Short-Time Frequency Transform (STFT)   A process of estimating the frequency response of a system from a collection of short Fourier transforms.

SINC   A function of the form $\sin(x)/x$ that represents an ideal Shannon interpolation filter.

Single Instruction Multiple Data (SIMD)   An architecture where individual processing elements perform the same instruction on many pieces of data.

Spectrum Analyzer   An instrument that displays the frequency-domain representation of a signal.

State Variable   Information-bearing variables residing in the shift registers of a digital filter.

State Variable Model   Defines a linear algebraic framework for the systematic representation and analysis of linear systems.

Stopband   The frequency range of a filter that is highly attenuated.

Sub-Band Filters   A part or portion of a larger filter solution.

Superposition   A property of a linear discrete or continuous-time system.

Superscalar   A CPU architecture implements a form of parallelism allowing the system as a whole to run faster than it would.

Time Division Multiplexing   A technique of packing a number of independent signal occupying nonoverlapping time intervals into a single signal stream.

Time Domain   A domain in which signal and systems are defined in terms of their time attributes.

Time Series   A continuous string of sample values.

Transducer   A transducer is a piece of equipment that converts a physical signal into an electrical signal.

Transfer Function   The ratio of the z-transform of an at-rest linear system's input to output z-transform.

Transpose Architecture   An architecture defined in terms of transpose operations applied to a different architecture.

Transversal filter   See finite impulse response (FIR) filter.

Truncated Fourier Transform Finite Impulse Response (FIR)   An FIR synthesis technique that approximates a desired frequency-domain response.

Twiddle Factor   The complex coefficients found in a fast Fourier transform (FFT).

Undersampling   A sampling rate below the Nyquist sample rate.

Very Long Instruction Word (VLIW)   Large-sized complex instructions encoded into one instruction.

von Neumann architecture   A traditional microprocessor architecture.

Warping   Nonlinear phase distortion introduce by the bilinear z-transform.

Wavelet   A means of representing or producing a signal in time and scale (frequency), which obeys a set of mathematical properties.

Window   A means of isolating, localizing, and modifying data over a finite interval to the exclusion of all time intervals.

Window Method Finite Impulse Response (FIR)   Technique of synthesizing an FIR filter in terms of a window function.

Worst Case   The maximum possible outcome.

z-Domain   The complex plane in which signals and system are represented by z-transforms.

Zero   An artifact (singularity) generated by the feed-forward path of a system.

Zero-Order Hold   A device or method that interprets a discrete-time signal as a piecewise constant analog signal.

Zero Phase Finite Impulse Response (FIR)   A noncausal FIR filter having a zero phase response.

z-Transform   A mathematical method of analyzing and representing discrete-time signals.

INDEX

Digital Filters: Principles and Applications with MATLAB, First Edition. Fred J. Taylor.
© 2012 by the Institute of Electrical and Electronics Engineers, Inc.
Published 2012 by John Wiley & Sons, Inc.